Liberty's Grid

Liberty's Grid

A Founding Father, a
Mathematical Dreamland,
and the Shaping of America

Amir Alexander

The University of Chicago Press
Chicago and London

The University of Chicago Press, Chicago 60637
The University of Chicago Press, Ltd., London
© 2024 by Amir R. Alexander
Published 2024
Printed in the United States of America

33 32 31 30 29 28 27 26 25 24   1 2 3 4 5

ISBN-13: 978-0-226-82072-9 (cloth)
ISBN-13: 978-0-226-82073-6 (e-book)
DOI: https://doi.org/10.7208/chicago/9780226820736.001.0001

Library of Congress Cataloging-in-Publication Data

Names: Alexander, Amir, author.
Title: Liberty's grid : a founding father, a mathematical dreamland,
   and the shaping of America / Amir Alexander.
Description: Chicago : The University of Chicago Press, 2024. | Includes
   bibliographical references and index.
Identifiers: LCCN 2023040934 | ISBN 9780226820729 (cloth) |
   ISBN 9780226820736 (ebook)
Subjects: LCSH: Jefferson, Thomas, 1743–1826. | Grids (Cartography)—
   United States. | Grid plans (City planning)—United States. | Mathematical
   geography. | Cartography—United States. | LCGFT: Altases.
Classification: LCC G1201.B7 A4 2024 | DDC 526.0973—dc23/eng/20231012
LC record available at https://lccn.loc.gov/2023040934

To Michal and Nitza

Americans do not read Descartes, but they respect his maxims.
—*Alexis de Tocqueville*

We rush like a comet into infinite space.
—*Fisher Ames*

We came here and created a blank slate. We birthed a nation from nothing.
—*Rick Santorum*

Contents

Introduction

About a mile from my house, surrounded by open space and hiking trails, is a craggy hill known locally as Castle Peak. Unlike its grander namesakes in Colorado and Idaho, "our" Castle Peak rises only six hundred feet above its surroundings and is too small to be considered a proper mountain. But in truth, it rather looks like one: its slopes rise dramatically on every side, their steepness increasing with altitude till they culminate in a rocky summit. It is our most notable landmark, our mini-Matterhorn, and it can be seen from almost every point in the neighborhood, peeking unexpectedly from behind a red roof or green palm tree. And at dusk, when viewed from just the right angle, the giant boulders that crown its peak do indeed resemble the turrets of a medieval castle, keeping watch over the quiet streets below.

On every New Year's Day for the past two decades, Bonnie and I set out to summit Castle Peak. The hike takes all of an hour each way, but it is not exactly a leisurely stroll: the path grows steeper and more challenging the higher we climb, and the final stretch involves some actual rock climbing. By the time we reach the top and scramble onto the highest boulders, we are usually out of breath, hot, and not a little sweaty. We take in the scenery and try to locate our home down below.

Castle Peak and El Escorpion park as seen from my neighborhood in West Hills, California. Photo courtesy of the author.

We think of what has happened in the year since we last ascended the peak and wonder where we will be when we next return.

We are not the first to appreciate the magic of this place. To the native people who lived here before Europeans arrived, these hills were sacred grounds. It is where the Chumash people of the coast and the Tongva people of the interior would come to trade, worship, and celebrate. They left their mark in pictographs of men in feathered headdresses that still line the walls of a cave not far from here, though its exact location is kept secret, for fear of vandalism. Looking west from the top of Castle Peak to the bare grassy hills of El Escorpion Park, it is easy to imagine their presence here. Not much, it seems, has changed. But turning west, one is confronted with the urban forest of Los Angeles's San Fernando Valley, stretching into the dusty haze as far as the eye can see. Everything, in fact, is different.

There is Victory Boulevard, running east in a perfectly straight line across the valley to the city of Burbank, more than twenty miles away. Sherman Way and Vanowen Boulevard, just as linear and as wide, run parallel to it without deviating for mile after mile after mile. Equally broad and straight cross streets intersect them from north to south, at precise right angles and regular intervals. The romantic-sounding Topanga Canyon Road is one; though named in the Tongva language, it cuts through the flatlands like a straight arrow, before winding its way into the Santa Monica Mountains and down to the Pacific Ocean. Parallel to it on the Valley floor is Owensmouth Avenue, its fantastic name commemorating the theft of the water of Owens River that made possible the settlement of this arid California valley.

This is what locals call "the Valley" and what millions around the world would easily recognize as the background to countless movies and television shows. It is a peculiar place, both famous and infamous, humdrum suburban and boisterously garish, and it has been my home for over twenty years. But when viewed from the heights of Castle Peak, it becomes clear that in one respect at least the San Fernando Valley is entirely typical: it is a thoroughly gridded landscape, just like hundreds of cities and thousands of small towns and neighborhoods across the United States. The street plan of the Valley is indeed the street plan of America: from New York and Miami in the east to Seattle and Los Angeles in the west, the streets of America offer a spectacle of rectilinear uniformity. Broad, arrow-straight avenues, regularly spaced and perfectly parallel to one another, are met at fixed intervals by equally straight and parallel streets that intersect them at precise right angles. The result is the ubiquitous and familiar urban grid, the hallmark of American cities. Now, rectilinear city streets were not, to be sure, invented in America.[1] But only in America are they so prevalent, so similar, and so open ended—unbounded and extending ever farther into the surrounding countryside. So much so that they seem like components of a single, all-encompassing urban grid stretching from New York to Chicago to Los Angeles.

And it is not only on their city streets that Americans show their preference for straight parallel lines and precise right angles. Indeed, anyone looking down on the western continental United States from the window of a commercial jetliner at thirty thousand feet will recognize the pattern. For hour after hour, all one sees is a blanket of square fields, oriented to the points of the compass and bounded by regularly spaced parallel lines that continue from horizon to horizon. On mile after country mile, one square field follows another in endless succession in every direction, overruling, skipping over, and ultimately ignoring mountains, valleys, rivers, creeks, and even cities. From end to end, the brown-and-green checkerboard covers a full two-thirds of the continental United States in a stunningly regular and monotonous pattern. It replaces a natural landscape with an abstract mathematical design, as precisely measured and outlined as a sheet of graph paper.

Just like rectilinear streets, the idea of dividing agricultural land in a rectilinear pattern did not originate in America. Scattered examples of square fields arranged in a checkerboard pattern can be found in northern Italy, the Dutch lowlands, and Japan, dating back to medieval and even ancient times.[2] But in all these cases the rectilinear pattern is local and clearly circumscribed, its scale and orientation conditioned by the peculiarities of the landscape. The grid that overlays the American West is something else: a single, unified, unlimited network of colossal size, oriented to the points of the compass, with no boundaries but those of the nation and the continent. Far from accommodating local conditions, it supersedes them, forcing every forest, stream, and hill to find its place within the mathematical pattern. It does not so much divide up space as redefine it: a rich and varied land of mountains, valleys, rivers, and forests, rich in wildlife and human history, is recast as a uniform and monotonous mathematical space.

In its bustling cities and in the open plains, the United States is inscribed with an endless succession of equally spaced parallels, intersected at right angles by equally spaced parallels. It seems like a mathematical, not a natural, landscape, and it is indeed so: For as we shall see, both in the agricultural heartland and in urban centers, locations are most often defined as they are in a Cartesian grid—by the intersection of two numbered rectilinear coordinates. Colossal in size, unbounded in extent, and unrelenting in its strict adherence to mathematical order, the grid is, in the words of one geographer, "mind-boggling."[3] It is also quintessentially American. Because, to put it simply, there is nothing like it in the world.[4]

Why did Americans choose to structure their newly colonized continent as an abstract mathematical landscape? We might, at first, be tempted to consider this a practical choice, a mere bureaucratic convenience whose only purpose is facilitating real estate transactions. And it is certainly true that early promoters of the grid often cited clearly defined boundaries and the easy sale of land as factors in the decision to carve out both city and country in a clear and simple manner.[5] But any attempt to explain rectilinear America as practical or utilitarian is bound to fail. From the beginning, the great grid was implemented in direct opposition to long-established traditions, and in the face of fierce resistance from some of the most powerful men in the land. Even George Washington himself staunchly opposed the plan, arguing that it would impede, rather than facilitate, settlement and expansion. Most critically, the establishment of a single grid on a continent-sized landmass was simply beyond the technical capabilities of surveyors when it was first proposed to Congress.[6] In the end, imposing a single Cartesian grid over the western United States and Alaska required a titanic multigenerational effort by a vast government bureaucracy created for that express purpose. Throughout, it was conducted at the frontiers of technical feasibility and lasted nearly two centuries. There was nothing simple or easy about it. There was nothing that could be considered "practical."

But if the grid was not a practical convenience, why was it created? A clue might be found in the identity of the person who first proposed and sponsored it, for the author of the Great American Grid was also the author of the Declaration of Independence. The man who insisted on carving an entire continent into tiny regular squares was also the man who boldly proclaimed that "all men are created equal" and "are endowed by their Creator with certain unalienable Rights." He was, of course, Thomas Jefferson. And while Jefferson's words

The gridded land: Southwest Iowa and Chicago as seen from space. Contains modified Copernicus Sentinel data, 2023, processed by Sentinel Hub.

in the nation's founding document have long been taken to express Americans' highest principles and idealized view of themselves, this book will show that much the same is true of his other creation—the meticulous rectilinear grid that overlays so much of the nation's territory. Written not on parchment but into the mountains, valleys, and plains of North America, the Great American Grid embodies an ideal of America as a land of unconstrained freedom and infinite opportunity. For the Great American Grid is not a utilitarian creation but an ideological one, presenting a distinct vision of the land and its people. And never for a moment, from Jefferson's day to ours, has it ceased disseminating that vision to the people of America.[7]

The notion that a featureless, uniform grid might be endowed with political or ideological meaning must strike us as surprising. Indeed, to modern mathematicians the Cartesian grid is nothing more than the implicit space in which their investigations take place. It makes possible the plotting of curves and the definition of mathematical objects, which may be of great interest. But it is of no significance in itself. And what is true of mathematicians utilizing the abstract grid is surely true of historians and social scientists contemplating the rectilinear pattern imprinted on the lands of the American West. For what could possibly be more neutral, more meaningless, than a pattern of evenly spaced, arrow-straight parallel lines, intersected at right angles by an identical set of evenly spaced, straight parallel lines? What could be emptier and, to put it bluntly, more boring than to have that pattern continue mile after mile in every direction, seemingly to eternity? Little wonder that of the many scholars who wrote about the American West, the vast majority all but ignored the grid—despite the fact that it is the land's most obvious, outstanding, and inescapable feature. Even those who did mention the grid almost uniformly considered it a functional convenience, the invention of bloodless bureaucrats, sometimes working hand in glove with avaricious speculators. In itself, the grid was of no significance.[8]

But things were not always so. There was a time when the mathematical grid, which we find so boring as to be invisible, was the focus of fierce debate that extended far beyond the bounds of mathematics.

For when it was first introduced in the late seventeenth century, the familiar rectilinear pattern stood for a radical new idea: that space could exist in and of itself, boundless and completely empty, irrespective of the objects that occupy it. The notion was key to Isaac Newton's *Principia Mathematica* of 1687, quite possibly the greatest scientific work in history. In Newton's system of the world, the sun, moon, planets, and comets move through absolute empty space, which itself is pervaded by the universal force of gravity. Turning to mathematics, Newton defined mathematical space in precisely the same terms: it too was a uniform, boundless, and infinite void containing disparate objects. The only difference between the two was that physical space contained physical objects, whereas mathematical space contained mathematical ones: in place of stars and planets, it contained curves and geometrical shapes, which could be defined, plotted, and manipulated within it. To mark the presence of absolute mathematical space, Newton made use of the familiar right-angled cross of the x and y coordinates and of the rectilinear grid they implicitly create. And while the use of these "Cartesian" coordinates ultimately became universal among mathematicians, it was Newton who first systematically marked them in his diagrams.[9]

Newton's conception of space, both physical and mathematical, seems to us all but self-evident. It is admittedly true that as a result of Albert Einstein's revolutionary work in the early twentieth century, physicists' conception of space today is very different from what it was for Newton in the seventeenth century. But the implications of Einstein's general theory of relativity are so at odds with our intuition and experience that they are all but meaningless to our lives.[10] For the most part, we conceive of space much as Newton did, and we live our lives within it. There is nothing surprising to us in Newtonian space.

But in the eighteenth century, Newtonian space, backed as it was by the most powerful scientific theory the world had ever seen, was felt as an intellectual earthquake. Loyal followers of Descartes, who insisted that space was indistinguishable from matter, argued that the concept of empty space was irrational and philosophically incoherent. Newton's rival, philosopher and mathematician Gottfried Wilhelm

Leibniz, charged that the concept of absolute space was an affront to religion. An infinite space that exists from eternity to eternity, independent of matter or anything else, he argued, was effectively another God. The Almighty, Leibniz argued, designed the world as a flawless mechanical clockwork, and only a "full" universe, in which matter pushed against matter with no gaps in between, would guarantee the undisturbed operation of the universal machine. To suggest, as Newton did, that God operated in empty, unconstrained space was to question God's wisdom and his ability to create a perfect machine.[11]

Newton rejected such criticisms and argued forcefully that the Cartesian concept of the plenum (i.e., all space filled with matter) leads inevitably to a dead end in science and to Godless determinism in religion.[12] Absolute space, argued Newton's spokesman, the Reverend Samuel Clarke, was not "another God" but an aspect of the divine he referred to as "God's sensorium." He then proceeded to skillfully turn the tables on Leibniz, who had argued that Newtonian absolute space was an affront to God's perfect wisdom. To the contrary, Clarke argued, it was Leibniz's deterministic plenum that was putting limits on God by denying him the ability to act freely in the world. Only absolute empty space, he insisted, allows room for God's unlimited power and unconstrained will.[13]

In the end, it was the philosopher John Locke, a friend and admirer of Newton, who drew the threads together. It is not only God's freedom that was made possible by Newtonian space, he argued, but man's freedom as well. Whereas the deterministic plenum of Descartes and Leibniz turns humans into something akin to mechanical automatons, Newton's unconstrained void allows humans to act freely and rationally. Absolute empty space, the space defined by the x and y coordinates and the Cartesian grid, had become for Locke the space of freedom.

And so, indeed, it would remain, throughout the age of Enlightenment and beyond. Human autonomy and freedom, cherished by philosophers such as Locke, Jean-Jacques Rousseau, and John Stuart Mill, required the empty space of Newtonian cosmology, a space in which anything was possible. Human and political rights, such as those enshrined by Jefferson in the Declaration of Independence, were also made possible by Newtonian space. It is that empty expanse that God, acting freely, endowed with the mysterious universal force of gravitation; and it is there that he freely endowed humans with the universal right to "life, liberty, and the pursuit of happiness." Such autonomy and freedom, for God as for man, would never have been possible in the deterministic plenum that made up the universe of Descartes and Leibniz.[14]

Locke, who was deeply involved in the establishment of British colonies, was quick to project the Newtonian notion of empty space onto the newly discovered lands overseas. Professing a doctrine that came to be known as "vacuum domicilium," he argued forcefully (albeit falsely) that America was an empty land in which English colonists could settle as they pleased. Following Newton's own practice of marking empty mathematical space with a Cartesian grid, he even proposed that American lands should be divided up in a rectilinear checkerboard pattern, a forerunner of what would become the great grid that covers the western United States.[15] Locke's plans for America were, to be sure, limited in scope and ultimately came to naught. But the same cannot be said of the designs of one of his greatest colonial admirers: Thomas Jefferson.

Jefferson's plan for western lands dates to the earliest days of the United States. The Peace of Paris, which in 1783 ended the Revolutionary War, handed the young republic nominal control of vast territories west of the Appalachian Mountains, and Jefferson was the man put in charge of deciding what to do about them. Jefferson, as it happens, was a mathematician of considerable talent and skill, and the solution he proposed was, accordingly, strictly mathematical. In successive reports issued by his congressional committees in 1783 and 1784, he outlined a seductively simple plan. Equally spaced parallel lines running from north to south would be marked throughout the entire territory, intersected at right angles by equally spaced parallel lines running from east to west. The resulting grid would divide the vast western landscape into a checkerboard pattern of perfectly square plots, oriented to the points of the compass. These would form the boundaries of new states, new counties, and ultimately individual plots to be sold to settlers. Finally, each square would be identified as the intersection of the x and y coordinates: an east–west coordinate (later called "range") and a north–south coordinate (later dubbed "township"). The western United States, in Jefferson's plan, was to become a single, gigantic Cartesian grid.

Jefferson's vision of a graph-paper West was both deceptively simple and strikingly audacious. Never before had anyone proposed imposing a unified mathematical scheme on a vast and varied terrain the size of a continent. But if there is one thing even more remarkable than the scheme itself, it is the fact that it succeeded. The landscape of the western United States today, though different in its details, is fundamentally just as Jefferson envisioned it: a single, massive, unified Cartesian grid, in which each square tract of land is defined as the intersection of two rectilinear coordinates. Over the course of two centuries, Jefferson's dream has become a reality.

Why did Jefferson seek to transform the rich and varied landscape of the North American continent into a monotonous and abstract mathematical dreamland? It was not, as some have argued, because the grid offered a practical solution to a pressing problem, such as the rapid distribution of land. It was, rather, because for him the Cartesian grid was not the meaningless pattern it appears to us but a scheme redolent with philosophical, political, and moral significance. For Jefferson, following Newton and Locke, gridded space was empty space, and empty space was where men could be free. It was, in other words, precisely what America was in Jefferson's mind. And if the real, actual America, with its diverse landscapes and rich human history, did not quite match his ideal of the blank-slate land, then it must be made to match it. By gridding the West, he sought to make it the true America, the land of his dreams.

Throughout his long public career, Jefferson never wavered from his vision of the American West as a vacant land of freedom. And though he never journeyed farther west than Virginia's own Shenandoah Valley, this absence of personal experience did not hinder—and likely aided—his expansive views on the West. For what he envisioned was not the actual rich and varied land, inhabited by a startling range of peoples and cultures, which stretches from the Appalachian Mountains to the Pacific Ocean. It was, to the contrary, an abstract empty space—uniform, undifferentiated, and ready to be exploited. By inscribing a single massive Cartesian grid onto the western United States, Jefferson sought to make it into the America of his vision: empty, passive, uniform, and awaiting the settlers who would use it to their best advantage.

Jefferson's grid was, and remains, a bold statement of what he believed America should be, a grand vision of the land and its people. For the land of the grid is the land of the American dream, an unbounded land of limitless opportunity, where people live freely, create their fortunes, and forge a nation. It was in that vacant and uniform mathematical terrain, Jefferson believed, that enterprising settlers, unconstrained by history, tradition, or geography, would build "an Empire of Liberty."

Liberty's Grid tells the story of the Great American Grid. It starts with an account of the grid's mathematical and ideological roots and continues with the early decades of its implementation on the western and urban landscapes. It then reaches forward to touch on the ways both the grid and its opponents have shaped America, in the late nineteenth century and even to this day. The emphasis throughout is on the power of both the grid and its counterpart, the anti-grid, to embody and disseminate their visions of America.

The book begins with a mathematical prologue, discussing the origins of the grid and its development in opposition to rival concepts of mathematical space. By positing the x and y coordinates, mathematicians would open up an infinite and empty space, in which every point is uniquely defined as the intersection of two rectilinear lines. It moves on to the British colonies in America, where Thomas Jefferson was developing his views on the land and its people. Though drawing on older traditions that viewed the New World as an empty expanse awaiting European settlers, Jefferson provided them with a powerful philosophical grounding, later combined with an aggressive expansionist policy. A mathematician of no mean ability, Jefferson identified his "Empire of Liberty" with gridded mathematical space and set about inscribing rectilinear coordinates on the western landscape.

The book continues to the implementation of Jefferson's radical plan for the West, from the early (and largely unsuccessful) attempts in Ohio in the 1780s, to the systematic rectilinear survey launched by Jared Mansfield, Jefferson's handpicked surveyor general, in the early 1800s. Over the following century and a half, 1.4 billion acres in the continental United States (including Alaska) were imprinted with a single immense Cartesian grid. And just as the western prairies were being inscribed with rectilinear lines, so were American cities. The most iconic of these is New York, where the natural landscape of the Island of Hills (also known as "Manhattan") was replaced by a flat, featureless mathematical grid, in which every point is defined as the intersection of two numbered arrow-straight streets.

Yet for all its grandeur and ambition, Jefferson's vision was not the only one seeking to inscribe itself on the American landscape. Not long after the rectilinear surveyors first set down their instruments on the western plains, they were challenged by those who viewed their mission with profound skepticism. The unchecked settlement of the West, the critics argued, brought about not only freedom and opportunity for settlers but also the destruction of the natural environment and the annihilation of the people who lived within it. Transcendentalists such as Henry David Thoreau, urban reformers such as Frederick Law Olmstead, and conservationists such as John Muir viewed the rectilinear terrain not as a land of freedom and possibility but as an oppressive artificial imposition, crushing to the human spirit. Let loose within a soulless mathematical grid, they believed, people are not free but lost, set on a path to social and moral degradation. The only cure for this midcentury malaise, they believed, was to check the spread of the Cartesian terrain by circumscribing it with naturalistic landscapes.[16]

The first major victory in the campaign against the grid came in the 1850s with the creation of "the central park"—a massive naturalistic expanse in the heart of Manhattan. It was a pointed rebuke to the rectilinear landscape, and it was not long before every American city was clamoring for its own refuge from the rigors of the grid. In time, the parks at the heart of gridded cities were joined by green naturalistic suburbs and leafy college campuses at their boundaries. There, curved streets and paths meandered along bubbly creeks and intersected at all angles except ninety degrees. Indeed, my own neighborhood, clinging to the edge of the San Fernando Valley grid, is just such a suburb—a landscape of winding roads, curved streets that end where they begin, and hidden cul-de-sacs, with nary a straight line or a right angle. And even as naturalistic suburbs and parks grew up within the gridded cities, natural landscape preserves began popping up in the rural landscape. With the creation of the National Park System and its attendant network of state parks, forests, and preserves, the "natural" vision of America came to challenge the open vistas of the rural western grid just as it did the enclosed city grids of Manhattan, Chicago, and San Francisco.

Shaped by competing and possibly irreconcilable visions, the American landscape has become a landscape of conflict. The Great American Grid is still there, complete with its Jeffersonian vision of America as a land of unconstrained freedom and limitless possibilities. But so is the anti-grid, warning that humans must walk humbly with nature, or they are surely doomed. In America, wherever there is one, there is also the other, the two clashing, encircling, and constraining each other, stalking each other across the land. Rigid rectilinear cities give birth to naturalistic parks at their center and curvilinear suburbs at their outskirts. The immense gridded West is interrupted repeatedly by protected natural wonders. On the one side are the streets of Manhattan and the cornfields of Kansas; on the other, the paths of Central Park and the cliffs of Yosemite Valley. And looking down from Castle Peak, on one side is the immaculate grid of the San Fernando Valley; on the other, the rolling open hills of El Escorpion Park. The warring visions of America are inscribed on the land and continue to exert their power on its people. On occasion, as at the Bundy Ranch in Nevada, the Malheur Natural Wildlife Refuge in Oregon, and the Standing Rock Sioux reservation in the Dakotas, they erupt in protest and violence. The fight rages on; the passions undimmed.

Finally, a personal note: I am, by training, a historian of science and mathematics and come as something of an outsider to the rich fields of Western and urban history. There are, inevitably, some risks involved, for a newcomer will be hard-pressed to match the breadth and depth of knowledge of scholars who have dedicated a lifetime to the field. Conscious of these dangers, I did my best to tread carefully, and I apologize in advance for any errors that may have crept into the narrative. They are entirely my own. And yet, along with the risks come opportunities as well: as an outsider, I bring to bear insights from my own specialty that may be unfamiliar to the field's resident scholars. By applying them to the new subject matter, I might add a fresh perspective, or even a new dimension, to historical questions that appear settled or unproblematic. This, I hope, is the case with this book. The massive grid that covers so much of the United States, both urban and rural, appears to both scholars and laymen as a meaningless bureaucratic convenience, so obvious and predictable as to be invisible. But viewed through the lens of historical mathematics, it becomes something very different: a radical and ambitious construct, imbued with an ideological vision and the power to implement it.

Liberty's Grid is the tale of the grid, the anti-grid that arose to confront it, and their battle over American space. Though reaching to the present, it begins as far back as the seventeenth century: it was then that a novel mathematical conception of space was conceived in northern Europe and then disseminated across the world. It begins, in other words, with the invention of a new kind of mathematics.

1

A Mathematical Prologue

The Stove-Heated Room

In the towns of southern Germany, winter had arrived early. Only a few months had passed since August 1619, when a new emperor was crowned in Frankfurt amid brilliant festivities that had lasted more than a month. Under the warm summer sun, the greatest princes of the realm, sparkling in their colorful attire and escorted by armed retinues, had gathered to celebrate the accession of Ferdinand II of the house of Habsburg to the throne of the Holy Roman Empire. But in the dreary frost of November, the brilliant festivities were but a distant memory. The lords and ladies had long departed, returning to the castles and palaces of their domains. Cold had set in, and frequent rains had turned the roads into rivers of mud, impassable to all but the most determined traveler. And as for the armed retinues, they had abandoned the parade grounds and moved into their winter barracks. They were preparing for war.

It was along one of those gloomy roads, not far from the free imperial city of Ulm, that a traveler came passing through on the tenth day of November 1619. Young and fit, with a well-groomed mustache and a trimmed beard, he spoke passable German, but with an accent that immediately disclosed that he was a Frenchman. His military bearing showed him to be a soldier, one of far too many who could be found crisscrossing German roads in those troubled days. For, mercifully unknown to the local residents, 1619 was the second year of what would become the longest and bloodiest of European religious conflicts. Over the course of the coming decades, the Thirty Years' War would lay waste to the prosperous towns and villages of Germany.

Such future devastation was but an ominous cloud on the horizon on the day when the young traveler stepped into the bustling inn of a local village. Accustomed as he was to wayfarers of all types, the innkeeper may well have given this visitor a second look. Though he was sturdy and of martial presence, his good manners and fine dress made clear that here was no ordinary soldier given to brawling, gambling, and whoring. This was a man of good breeding. Unusually, the stranger requested a room for himself alone, and as he was willing to pay, the innkeeper showed him to a small but cozy room heated by a wood-burning stove. In response to a query, the traveler introduced himself: he was René Descartes, gentleman of the town of La Haye in Brittany.[1]

In 1619, Descartes was only twenty-three years old but had already traveled more, seen more, and experienced more than most of his contemporaries would in a lifetime. Born in 1596 to an ancient and respected family, he was sent at the age of eleven to study at the famous Jesuit College at La Flèche. There Descartes received the best education available in his day, including not only theology, as would be expected in a Jesuit institution, but also what was known as the humanist studies: ancient and modern languages, rhetoric, the classical authors, Aristotelian philosophy, and—critically for his future endeavors—a solid grounding in mathematics.

After seven years of study, Descartes left the Jesuit College as a cultured and well-educated young man but also, as it turned out, a profoundly disillusioned one. Having studied the works of the greatest philosophers who had ever lived, he felt he still did not know anything for certain. Those wise and learned men, it seemed, could not agree even on the simplest things, leaving him to wonder whether he had actually gained any knowledge in all his years of study. The only reasonable course, he concluded, was to doubt everything he had learned, no matter the plausibility or the source. "Considering how many diverse opinions learned men may maintain on a single question—even though it is impossible for more than one to be true," he wrote in the *Discourse on the Method* years later, "I held as well-nigh false everything that was probable."[2]

The crisis of confidence about truth and knowledge, which Descartes was confronting in himself, was not his alone. It was, rather, emblematic of his age. For a thousand years, the church had reigned supreme in the minds of Europeans, its authority unchallenged not just in religious matters but also in areas of learning we would consider entirely secular. If one wanted to know anything about the workings of the world, animals, the motions of the planets, or musical theory, one need only turn to the writings of the great masters authorized by the church. In theology, those were the church fathers and medieval scholars such as Peter Lombard and Thomas Aquinas. In astronomy and geography, it was Ptolemy of Alexandria; in geometry, Euclid; and so on. And the greatest authority of all, in fields ranging from logic to metaphysics to natural philosophy, was the ancient Greek sage Aristotle, who was known simply as "the Philosopher." It was a comforting, fixed, and unchanging world, in which everything that could be known was already contained in the works of the great masters. Every question had an answer, which was authorized by the church and therefore inescapably true. And if you did not know what the answer was, you certainly knew where to find it.

Yet this familiar world, in which everything that could be known was, came crashing down around the turn of the sixteenth century. First came the discovery of the Americas, which was no discovery to their millions of inhabitants but which nevertheless came as a shock to Europeans. If the greatest authorities, including Aristotle, were ignorant of an entire New World, what else were they ignorant about? And could they truly be trusted? Some decades later, Nicolaus Copernicus, a highly respected astronomer, was promoting the idea that the Earth was not stationary at the center of the cosmos but was racing around the sun at dizzying speed while spinning on its own axis. Most astronomers of the day rejected Copernicus's view, and yet a nagging seed of doubt had been planted: to most people, the knowledge that the Earth is solid and unmoving beneath their feet was the most basic, irrefutable truth of daily life. If one could not be sure of that, what could one be sure of?[3]

But the most crippling blow to the fixed and known medieval world came from the very heart of the institution that supported it—the church itself. On All Saints' Day, October 31, 1517, an obscure Augustinian monk and professor named Martin Luther posted a list of ninety-five theses on the door of the Castle Church in Wittenberg, Germany, challenging core church doctrines. Luther, in all likelihood, did not mean to bring down the church but merely to cleanse it of error and corruption, but things turned out differently. His teachings spread like wildfire, first in Germany and then beyond, and he soon found himself the leader of a new "Protestant" religion that rivaled the Roman Church and condemned it as the incarnation of all evil. Soon, other reformers joined in, each claiming to be the bearer of true Christianity: Huldrych Zwingli in Zurich, Martin Bucer in Strasbourg, John Calvin in Geneva, and many more. The single unified voice of the Roman Church had splintered into a cacophony of competing voices, each insisting that it alone spoke the truth and that all others were charlatans, impostors, or worse. In such a world, how could one possibly know whom to trust or how to tell truth from error? In sixteenth-century Europe, the very possibility of obtaining true and certain knowledge was cast into doubt.

And yet Descartes was not discouraged. On leaving La Flèche, he decided to look beyond the learned books he had mastered under the Jesuits to knowledge he could trust. He spent several years, as he later recalled, "travelling, visiting courts and armies, mixing with people of diverse temperaments and ranks," and in 1618, with clouds of war gathering all across Europe, he enlisted in the army of Prince Maurice of Nassau. The following year, while on his way to join the forces of Maximilian, Duke of Bavaria, he stopped in Ulm, where as usual he engaged with local scholars and mathematicians. By the time he resumed his journey to Munich, Maximilian's capital, winter was setting in, making travel slow and often impossible.

And so it was that on November 10, 1619, Descartes found himself in a rustic inn on the road from Ulm to Munich. When shown to his quarters, he did something that must undoubtedly have appeared curious to the innkeeper, used as he was to boisterous and unruly soldiers: he closed the door and did not reemerge for a full day and night. Alone in that room, he wrote, years later, "Finding no conversation to divert me, and fortunately having no cares or passions to trouble me, I stayed all day shut up in a stove-heated room, where I was completely free to converse with myself about my own thoughts."[4]

As Descartes tells it, the roughly twenty-four hours he spent in this stove-heated room transformed him from a restless young man into a systematic philosopher who would upend the foundations of Western thought. Over the next twenty-five years, he would go on to publish a series of works that would become landmarks of Western philosophy: *Rules for the Direction of the Mind* (1628), *The World* (1633), *Discourse on the Method* (1637), *Meditations on First Philosophy* (1641), and *Principles of Philosophy* (1644). But as Descartes saw it, the roots of all of these works went back to the day he had spent holed up in the south German inn. It is difficult, to be sure, to say with confidence which ideas originated precisely on that day and which were added and elaborated later on. But Descartes's thought, perhaps even more than those of other leading philosophers, has an unmistakable unity to it, which holds his entire body of work together from beginning to end. And since Descartes considered his meditations at the inn to be the spring from which all his later thought flowed, it seems fair to follow his lead and assume that his core ideas originated in the stove-heated room on the road from Ulm and Munich.

A Mathematician's Universe

Years later, Descartes gave a detailed reconstruction of his meditations that day: How he began by banishing from his mind everything that contained the least taint of uncertainty. How in the end he was left with only a single truth that withstood the onslaught of doubt: that he is a thinking being who exists. How from his idea of perfection it followed that God necessarily exists, and from God's perfection it followed that he could not be a deceiver. Hence, Descartes concluded, ideas that were clear and distinct in his mind, so much so that he could not shake the conviction that they were true, must indeed be so. To suggest otherwise would be to charge a perfect God with deception. Finally safe in the knowledge that rigorous, clear, and distinct reasoning must necessarily lead to the truth, Descartes proceeded to build up the world from scratch.[5]

What is the world that we live in? Through clear and distinct reasoning, Descartes concluded that it is composed exclusively of matter and that matter is synonymous with extension in space: matter is identical with the space it occupies, and any defined space is, by definition, filled with matter. This has several crucial consequences: For one thing, a vacuum—or empty space—is impossible. For another, space and matter are uniform and extend indefinitely in every direction. Finally, and most importantly for us—space is mathematical, and everything within it—every location, every object, every shape, and every movement—can be precisely described through mathematics.[6]

To this basic scheme, Descartes added many details. He concluded, for example, that the only thing in the world apart from matter is motion and that everything can therefore be explained as matter in motion; that there are three types of "elements" in the world, differing in the size of their particles, which fill up all of space; that the natural motion in the world is circular; that the planets are carried around the sun in vortices of matter composed of the "second element"; that other stars are "suns" with their own planets; that animals are machines that only imitate life; and much, much more. Most of these ideas, for all their ingenuity, have long since been discarded by scientists who came up with far more sophisticated and precise descriptions of the natural world. But the idea that the space of the universe is uniform, indefinitely extended, and mathematical has stood the test of time. It may well be the simplest element in Descartes's overall scheme; but it is also the most revolutionary.

To understand just how radical Descartes's universe appeared to contemporaries, consider the universe it replaced: the classical Aristotelian cosmos, as it was conceived in ancient Greece two millennia before. In the early 1600s, when Descartes was developing his system, this was still the dominant picture of the world, the one taught at schools and universities. In Aristotle's scheme, the world, known by the Greeks as the "cosmos," was made up of a series of concentric and perfectly spherical shells. At the center was the sphere of the Earth, immobile and immovable since the beginning of time and containing all manner of animate and inanimate matter. The rest of the spheres were made of a hard crystalline substance, each revolving around the Earth carrying a different heavenly body—from the moon, closest by, to Saturn, farthest away.[7] The outermost sphere, known as the firmament, carried all the fixed stars together around the Earth once a day. And that is where the cosmos ended: beyond the firmament there was nothing, according to Aristotle. Not even empty space.

In nearly every single facet, the traditional Aristotelian cosmos was the opposite of the universe Descartes was proposing to replace it with. The traditional cosmos was finite; Descartes's universe was boundless (though not necessarily infinite). The cosmos had a center and external boundaries; the Cartesian world had none, as all points in its limitless space were the same. For Aristotle, *up* meant "away from the center of the cosmos" and *down* meant "closer to the center of the cosmos," and whereas "light" substances strove upward, "heavy" substances strove downward. For Descartes, in contrast, such distinctions were meaningless: all directions in uniform space were qualitatively the same. In the Aristotelian cosmos, different laws applied in different places, and places were qualitatively different from one another. The heavens, for instance, were the realm of perfection, whereas the sublunar sphere was the realm of change. But in Descartes's world, the same rational principles applied everywhere. The laws of nature were universal.

And then there is this: Descartes's universe was mathematical; Aristotle's universe was not. Mathematics, to Aristotle, was simply too precise, too certain, and in fact too perfect to apply to the irregular world we see around us. It might, perhaps, be applicable to describing the perfectly circular and regular motions of the heavens. But for describing and explaining the Aristotelian physical world, mathematics was useless.[8] The opposite was true for Descartes, in whose universe mathematics reigned. All objects in the Cartesian universe were defined by extension—that is, by their precise geometrical shape. Their motions, straight or circular, were geometrically described, and their interactions were accounted for through precise mathematical laws of motion. "The only principles that I accept, or require, in physics," he wrote, "are those of geometry and pure mathematics; these principles explain all natural phenomena, and enable us to provide quite certain demonstrations regarding them."[9]

We may think what we may of the rigor of Descartes's reasoning: of his all-encompassing doubt, his proof of God's existence, his reliance on clear and distinct ideas, and his conviction—so central to his system—that God is not a deceiver. Many a modern reader would find at least some of the steps in his chain of reasoning questionable, or perhaps worse. Armed with the scientific knowledge of the past four centuries, we may be dismissive of his physical theories, his denial of the possibility of a vacuum, his belief in three different elements that fill up space, and his insistence that vortices carry the planets along. We might question his famous dualism of mind and body or consider his view of animals as machines outright appalling. Yet we cannot deny that by all these questionable speculations, Descartes, for the first time, created something momentous: the modern universe.

For Descartes's universe is, in most respects, fundamentally our own. As we all learn in school, we live on a planet that moves through space, which extends without limit in all directions. This space has no center and no boundary, no "up" and no "down." It is homogeneous; it is everywhere the same, and the same principles that apply in our surroundings also apply in every corner of the universe. Even Descartes's contention that the fixed stars are suns like our own with their own planets has turned out to be not far off the mark. And most crucially, our universe is mathematical, as every object, location, mass, and motion in it can be described through precise mathematical equations. In all these respects, our universe differs radically from the Aristotelian cosmos, but for the very same reasons, it is also the direct descendant of the universe hatched in Descartes's mind. Our universe, it is fair to say, had its origins in one man's search for certainty in a world where certainty seemed lost forever.

The King of Empty Space

Descartes's universe did deviate from our own in one crucial aspect: its space was completely full, allowing no gaps of emptiness to exist between different bodies. This was obvious to Descartes, since matter for him was nothing but spatial extension, and therefore any space was by definition filled with matter. But others were not convinced. His older contemporary Giordano Bruno (1548–1600), the Italian polymath who was burned at the stake in Rome, believed that the universe was an infinite void sprinkled throughout with stars. The great Galileo, for his part, offered no opinion on the existence of a void between the planets but suggested that minuscule voids do, in fact, exist inside matter.[10]

But it was Isaac Newton's (1643–1727) theory of the universe that critically revised Descartes's views, giving our world the shape we are most familiar with today.[11] Newton's work spanned the range of scientific fields known in his time, and in field after field, his contributions were not just important but revolutionary. In 1666, while on leave from plague-stricken Cambridge, he developed the infinitesimal calculus; demonstrated that white light was, in fact, a composite of a range of different colors; derived the formula for calculating centripetal force $(F = m\frac{v^2}{r})$; and calculated the pull of the Earth's gravity on the moon, leading to the "inverse square law."[12] In the *Principia*, two decades later, he recast the laws of mechanics into three simple mathematical laws, giving them the shape they would largely retain to this day. Then, after positing that any two bodies attract each other in direct proportion to their mass and in inverse proportion to the square of their distance, he used his new laws to derive the motions of the heavenly bodies with a precision that was previously unimaginable.

That Descartes's writings made a deep impression on the young Isaac Newton, there is no doubt. He fully shared the Frenchman's con-

tempt for the Aristotelian canon that formed the backbone of the official curriculum in Cambridge, just as it had at La Flèche half a century
before. He was profoundly impressed with Descartes's trust in rigorous reasoning, though over the years he learned to balance it with
an equally strong emphasis on empirical knowledge. He wholeheartedly endorsed Descartes's boundless and homogenous mathematical universe, in which the laws of nature apply equally everywhere.
And he practically devoured Descartes's mathematical works, and in
particular his *Géométrie*, which supplied the mathematical tools that
underpinned his world. Indeed, the *Géométrie* is among the very first
books that we know Newton purchased not long after his arrival at
Cambridge.[13]

Yet Descartes's insistence that matter is synonymous with extension in space, and that consequently the space of the universe is entirely filled with matter, was completely unpalatable to Newton. Part
of his objection was scientific: Descartes had argued, for example,
that in his "full" universe the planets were carried around the sun
in a vortex of matter, but this, to Newton, hardly made sense. How,
he asked, does one derive the complex elliptical motions of varying
speed, precisely described in Kepler's laws of planetary motion, from
the simple circular motion of a vortex? And how can one possibly explain the extremely eccentric orbits of comets through such a mechanism, especially since the paths of those comets cut right through the
supposed vortices of the planets? The fact that the planets and comets
proceed on their courses without being in the least perturbed by one
another's presence was, to Newton, a strong indication that they were
moving not in a turbulent plenum of matter but rather in completely
empty space.[14]

But there were also other reasons that led Newton to reject Descartes's theory of a "full" universe. Today, we would call those reasons "religious" and sharply distinguish them from his "scientific"
reasoning, but Newton did not see things that way. "And thus much
concerning God," he wrote at the end of a long theological discussion
in a chapter of the second edition of the *Principia*, known as the "General Scholium," before adding, "to discourse of whom . . . does cer-

tainly belong to Natural Philosophy."[15] The problem with Descartes's universe, for Newton and other English critics, was simple: it left no room for God.

The universe, as Descartes described it, was entirely packed with matter, leaving no gaps or empty spaces. Since all objects touched other objects, the interactions between them were entirely mechanical—one object physically pushed another object, which in turn pushed against a third, and so on. The entire universe, in effect, operated as one astoundingly complex mechanical clockwork. This, to Descartes, was a great advantage of his scheme: a universe that operated entirely by the familiar means of direct mechanical impact required no divine interventions, no magical forces, and no mysterious spirits. It was, he and his followers argued, entirely self-sustaining, rational, and knowable through his method of clear and distinct reasoning.

But what seemed to Descartes to be a triumph of reason over occult superstition appeared very differently to his critics. To them, a world without gaps, in which every object is fitted snugly against others and whose motion is determined by their direct pressure, leaves no room for God's action in the world. If the universe operates as a perfect clockwork mechanism, then where will God exert his power? The Cartesian world was, they argued, too full and too perfect, governed by the cold rational principles of mechanics rather than by the benevolent providence of God.[16]

All of these issues weighed heavily on Newton's mind when he created his own picture of the world. He wanted to retain key features of the Cartesian world—its boundlessness, homogeneity, and rationality—while at the same time addressing those aspects that he found unacceptable. In what kind of world would the planets follow Kepler's laws of planetary motion and the vortices of the planets and comets not interfere with one another? Even more importantly, what kind of world would be rational and even mathematical and yet allow for God's presence and intervention? Newton believed he had the answer. Descartes's key error, in his opinion, was that he identified matter with extension, failing to distinguish between space and the matter that occupied it. Newton would not repeat this mistake: "Absolute space," he wrote at the beginning of the *Principia*, "is its own nature, without relation to anything external, remains always similar and immovable."[17] This means that space exists in itself, absolutely, regardless of whether it is occupied by a material object. And if space exists in itself, then a vacuum, or empty space, can exist as well. It is therefore entirely plausible that objects in the universe, and in particular the heavenly bodies, are not interconnected through a continuous ocean of matter; rather, they are separated from one another by a void—absolutely empty space.[18]

To a follower of Descartes, this was absurd: if objects are separated by a void, how do they interact with one another, and how does the great cosmic machine operate? Newton answered with a boldly speculative leap: bodies, he declared, do not need to touch one another to interact because they are connected by what we call today "universal gravitation," which pulls them toward one another. This force, Newton insisted, can be measured and described with mathematical precision, even though its source and nature remain unknown. Consequently, bodies can affect one another even when they do not physically touch. Cartesians, unsurprisingly, were not mollified: This "gravity," they argued, was nothing but a mysterious magical force, the product of Newton's fevered imagination, and the farthest thing imaginable from a "clear and distinct idea." Introducing such forces into the world, they argued, would be a colossal setback for natural philosophy (as science was then called). Just when knowledge of the world was emerging from the darkness of superstition to the light of reason, it was being thrust back into the shadows by Newton's inexplicable force of "gravity."

Newton and his associates, for their part, were unmoved. Far from being a magical or "occult" force, gravity was an undeniable empirical fact, measurable and answering to precise mathematical laws: every body in the universe attracts each and every other body in a force directly proportionate to their mass and inversely proportionate to the square of their distance. The fact that the source and manner of operation of this force were unknown was interesting but in no way disqualified it from being a key part of a rational system of the world. This was especially the case because the principle of universal gravitation possessed enormous, and in fact unprecedented, explanatory power.

By assuming universal gravitation, Newton was able to account for everything from the drop of a stone on Earth to the motions of planets, moons, and comets. And he could do it all with a degree of mathematical precision that was literally unheard of. The Cartesians might denounce the mysterious power of gravity all they wished and insist on the methodological purity of relying solely on clear and distinct ideas. But as long as Newton's system explained far more and far better than Descartes's system ever could, their complaints mattered little. The scholars of the age, even those who were troubled by the enigmatic nature of gravity, simply refused to discard what was plausibly the greatest scientific work the world had ever seen simply because it did not conform to Cartesian orthodoxy.

And this was not all. With one stroke, Newton's system eliminated the difficulties he and many of his contemporaries had with the Cartesian worldview. The scientific question of how planets that were carried along on vortices of matter would move in elliptical paths with varying speed simply disappeared once it was acknowledged that the heavenly bodies move through a void. So did the question of why the vortices of the comets and the planets did not interfere with one another. Instead, the heavenly bodies were held in their paths by a combination of their inertial motion and the gravitational pull of the sun and Earth, following orbits that could be mathematically calculated with immense precision.

Much the same was true of the religious objections to the Cartesian universe, which Newton shared with many of his countrymen. For if Descartes's world was a perfect mechanism that left no room for divine action, this was not the case in Newton's world. In the emptiness of space, bodies do not press against one another or push one another in a prescribed direction. Movement must originate from a nonmechanical source, which in Newton's world was, famously, gravity. It was possible that gravity had a natural source, which, although unknown, may someday be revealed. But it was also possible that the cause was supernatural and originated directly from God's will.[19]

Newton himself was noncommittal on the issue, famously stating that on the question of the cause of gravity, "I frame no hypotheses."[20] But he did leave clues to his thinking: In a January 1692 letter to his admirer the Reverend Richard Bentley, he wrote, "Gravity must be caused by an agent acting constantly according to certain laws, but whether that agent be material or immaterial I have left to the consideration of my readers."[21] When Bentley, taking the hint, declared that the source of gravity was indeed God, Newton did not object.[22] More explicitly, his friend, the mathematician David Gregory, had this to say after a conversation with Newton: *What cause did the ancients assign of Gravity. He believes that he reckoned God the cause of it, nothing els.*"[23]

Newton had turned the tables on Descartes. For the Frenchman and his followers, the fact that the causes of gravity were shrouded in mystery was a catastrophic flaw that should have disqualified the concept from being considered in natural philosophy. For Newton, it was the opposite: the very fact that gravity was not understood yet was everywhere present strongly suggested that it was nothing less than the direct intervention of God. Descartes sought to outlaw mysteries from the world, considering them irrational and what we would call "unscientific." Newton, in contrast, reveled in the unknown because it gave an opening to God's presence in the universe. He had found the perfect way to combine a rational universe, one that could be studied and understood by humans, with a God that was all-powerful and ever present in the world. The mathematical principles that governed the action of gravity demonstrated that the world was comprehensible—at least up to a point. The enduring mystery of gravity itself, however, was a sign that God was omnipresent and had not abandoned his world.[24]

Even the existence of the void was, to Newton, evidence of the ubiquitous presence of God. To the question of "*what the space that is empty of body is filled with,*" Gregory reports Newton's response as follows: "The plain truth is, that he believes God to be omnipresent in the literal sense; . . . for he supposes that as God is present in space where there is no body, he is present in space where a body is also present."[25] God, in other words, pervades absolute space, and as we move through space, we are literally moving "in" God. By transforming Descartes's "full" world into one dominated by emptiness and interconnected through gravity, Newton, in his own eyes, had done more than improved on a scientific theory: he had made the world godly again.

By the turn of the eighteenth century, thanks to its immense scientific power and also, in some circles, thanks to its religious significance, Newton's universe became the standard picture of the world. This, quite naturally, happened first among scholars, but it was not long before the picture spread far and wide among the broader educated segments of the European populace. Like Descartes's universe, it was homogenous, uniform, and mathematical, and it extended without limit in every direction. But unlike it, Newton's universe was actually infinite, not just "indefinitely extended." It existed within absolute immovable space, the heavenly bodies were separated by a void, and they interacted with one another not through direct impact but through the universal force of gravity. It was a simple and elegant picture, which Newton himself called "this most beautiful system of the sun, planets and comets" and which would remain westerners' prevailing view of the universe for more than two centuries. And to most people, even after Einstein's critical revisions, it still remains so today.[26]

Descartes Invents a New Geometry

For both Descartes and Newton, the world and everything within it inhabited an unbounded, homogenous mathematical space. But if this was so, how could the world be described? The old Aristotelian picture—which distinguished between the sublunar world and the heavens in a finite cosmos and in which every substance had its "natural" place that it sought to inhabit—clearly would not do. The obvious tool for describing this new kind of universe was the ancient science of geometry, as codified two millennia earlier in Euclid's *Elements*.[27]

Through the practice of rigorous and systematic reasoning, Euclid had established precise relations between geometrical objects and proved them to be universal and true beyond a shadow of a doubt. He proved, for example, that the sum of the angles of a triangle was equal to two right angles (or, as we would say, 180 degrees) and that this was true for any triangle that ever was or could be. Or similarly, he proved the Pythagorean theorem, which states that for a right triangle, the sum of the squares of the sides is equal to the square of the hypotenuse. Moving step by step from such relatively simple theorems to ever more complex truths, Euclid created an entire world of lines and triangles, circles, squares, and parallelograms, each related to other objects in ways that are necessary and irrefutably true. As the most rigorous science of its day, it seemed perfectly suited for uncovering the hidden mathematical structure of the world. The great Galileo spoke for many of that generation when he declared that the world is a book written in "triangles, circles, and other geometrical figures."[28]

Yet Euclidean geometry also suffered from severe limitations when it came to describing the world. Having remained almost unchanged for nearly two thousand years, it was a "completed" science whose results were undoubtedly true but were also fixed and held no surprises. This was useful as long as the figures one encountered in the world were indeed Euclid's own circles and regular polygons or if one managed to recast the world in ways that matched these standard figures. Galileo, for one, proved a master at doing just that, ingeniously pushing the limits of what could be described by standard Euclidean methods. But what if the figures one encountered were not accounted for in the geometrical canon? What if they involved change and bodies in motion, concepts that were alien to the Euclidean canon, which described only steady immutable states? What if they involved figures and shapes that could not be constructed, as the canon required, by compass and straightedge? For all those cases, Euclid and his later interpreters had no answer. To them, figures that did not accord with the standards of the canon were illegitimate and should be excluded from geometry.

In the end, it turned out that Euclidean geometry, for all its beauty and rigor, was simply too unwieldy to describe the complex mathematical worlds that were proposed by Descartes and others. Its methods of proof were powerful, but they were also difficult and cumbersome, requiring enormous ingenuity in each and every case. And in all too many cases, even the greatest ingenuity could not overcome the method's inherent limitations. Euclidean geometry indeed described a world—a rigorous and rational world made up of triangles, circles, and polygons. But though similar in some ways to our own, it was also obstinately different from it, and the gap between the two ultimately could not be bridged. To describe our own world, Descartes and others discovered, one must use a different method and approach. But what method would that be?

Descartes believed he had the answer. In 1637, along with the *Discourse on the Method*, he published three appendixes, each longer than the *Discourse* itself, intended to demonstrate the power of his method for correct reasoning. Two of them dealt with straightforward scientific topics: the *Optics* pertained to the laws of vision, and the *Meteorology* deciphered the mystery of the rainbow. The third, however, dealt with mathematics and offered a radical revision of geometry that would make it more flexible and more powerful than Euclid's. His new method, Descartes boldly proclaimed in the book's opening sentence, would solve "any problem in geometry." It was called *La Géométrie*, or simply *Geometry* in English, and it would become one of the founding texts of modern mathematics.[29]

Most people who have heard of the *Geometry* have likely been told that it contains the core principles of the modern field of analytic geometry. There is some truth to this assessment, but anyone approaching the text expecting to find the familiar themes they remember from high school algebra is in for a surprise. For one thing, despite their name, there is not a trace of "Cartesian coordinates" in Descartes's masterpiece. In fact there are no coordinates of any type in the *Geometry* (though in some places Descartes does make use of a single ordinate or x axis). There is, furthermore, no effort to draw the curves of new equations, a task that is at the heart of analytic geometry as we know it today. What one can find in the *Geometry*, however, is a systematic method for resolving geometrical problems through the use of algebraic equations, in a manner that would be completely unacceptable to practitioners of traditional Euclidean geometry. And there is also this: in the *Geometry* Descartes introduces the world to x and y, the standard variables of modern equations.[30]

To understand what Descartes was trying to accomplish in the *Geometry*, consider an example. In book 2, Descartes poses a question that is a special case of the more difficult ancient problem known as the "Pappus problem." Suppose there are four parallel lines at equal distances from one another, *FG*, *DE*, *BA*, and *HI*, and a fifth line, *GI*, perpendicular to them, as in the figure below.

What is the locus of all points *C* such that

$$CF \times CD \times CH - CB \times CM \times AI? \quad (1)$$

To resolve this, Descartes proceeded by analysis, meaning that he supposed that he had found a point *C* that answered the conditions, and then worked backward to discover which point it was. He labeled the distance *CM* as x, the distance *CB* as y, and *AI* as the fixed quantity a, which is the fixed distance between the regularly spaced

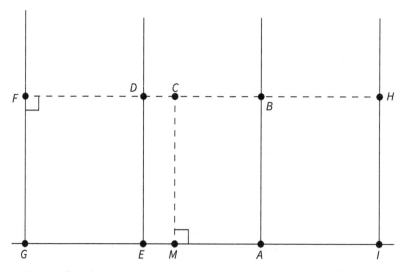

Diagram based on Descartes, *Geometry*. Descartes, *Geometry*, 336. Versions of the diagram can also be found in Boyer, *History of Analytic Geometry*, 87, and Grabiner, "Descartes and Problem-Solving," 89.

parallels. It follows that $CD = a - y$ and $CF = a + (a - y) = 2a - y$ and $CH = a + y$. By placing these equivalences in the above equation, the condition becomes

$$(2a - y)(a - y)(a + y) = y \times x \times a \quad (2)$$

which yields

$$y^3 - 2ay^2 - a^2y + 2a^3 = axy \quad (3)$$

For a modern mathematician, this equation would have been the solution to the problem and the end of the story. But not so for Descartes and his contemporaries. As they saw it, since the problem was posed geometrically, it must also be resolved geometrically. The question, accordingly, was this: What is the locus of all the points C whose distance x from the line GI and distance y from the line AB answers to equation (3)?

Since (3) is a cubic equation, Descartes knew that the locus of all the points was not one of the classical conic sections—the circle, ellipse, parabola, and hyperbola—which all answer to second-degree equations. Although he could not prove it, he strongly (and correctly!) suspected that whatever curve this was, it could not be constructed using only the traditionally sanctioned tools of compass and straightedge. Yet Descartes never doubted that to solve the problem, a geometrical curve must be constructed, meaning that it must be actually, technically drawn. His solution was to expand the range of allowable tools beyond the traditional ones, permitting more complex arrangements of two or more straightedges, which together could graph new classes of curves. Using these, he was able to construct the required cubic parabola. QED.[31]

With a boldness born of supreme self-confidence, Descartes was blasting through the rules and restrictions of Euclidean geometry, which for two thousand years had been the epitome of scientific knowledge. First there was Descartes's reliance on the method of analysis. Ancient geometers were well familiar with the analytic approach, but they considered analysis a flawed form of reasoning that could easily lead to error. Any result acquired by analysis, they insisted, must then be recast in the traditional synthetic mold—by arguing directly, step by step, from the stated assumptions to the proven solution. Descartes, however, had no such qualms. The analytic method, he argued, was perfectly rigorous and reliable in itself, not to mention more general and easier to use than the cumbersome synthetic method. And although Descartes did not himself coin the term *analytic geometry*, the practice of systematically applying analysis to geometrical problems can rightly be traced to him.

Then there is Descartes's willingness to mix geometrical problems with algebraic equations, in direct violation of the canons of Euclidean geometry. As anyone trained in mathematics knew, geometry was the science of magnitude, whereas arithmetic—and by extension algebra—was the science of quantity. Mixing the two was bound to lead to error. It was a time-honored ban, based on sound rational foundations, and yet Descartes violated it without as much as a nod toward the ancient separation of the sciences of magnitude and quantity. He did so believing that his approach fully accorded with the equally stringent principles of reason he laid out in the *Discourse*. And he did so, most of all, because it made his version of geometry that much more powerful.

Finally, Descartes rejected the traditional limitation on the use of geometrical tools, which proscribed the use of any instruments except compass and straightedge. Ancient geometers knew that some geometrical curves could only be produced through the use of more complex instruments. But such curves, termed "mechanical," were considered distinctly second class and were not part of the pure and certain world of geometrical objects. Descartes, however, would have nothing to do with the distinction between geometrical and mechanical curves.[32] If we accept the compass and straightedge (or "ruler") as legitimate tools, he argued, we must accept more complex ones as well, in which "two or more lines can be moved, one upon the other, determining by their intersection other curves."[33] He then made extensive use of these unconventional tools in constructing complex curves.

For the first time in two millennia, Descartes had produced a new kind of geometry. Elegant and rigorous though it was, the old Euclidean geometry focused on the creation of its own perfect world of absolute certainty, and trying to apply it to our own world was difficult and often simply impossible. Descartes, who insisted that everything in the world could and should be described mathematically, needed a more powerful, flexible, and far-reaching tool. In *The Geometry*, he produced precisely that.

In two significant ways, however, Descartes—despite his bold iconoclasm—remained true to the Euclidean tradition. First was the fact that Descartes insisted that the solutions to his equations must be properly constructed (although not necessarily by compass and straightedge alone) geometrical curves. Some later mathematicians, particularly in the eighteenth century, would dispense with geometrical representations, arguing that algebraic equations are solutions in themselves. For Descartes, however, such an approach would not do: his ultimate goal was to mathematically describe the world, which means to mathematically recreate all the complex shapes and figures within it. To stop short of this, and settle for abstract algebraic equations, would seem to him entirely pointless.

But Descartes remained true to the classical tradition in another sense, in the way he conceived of the most basic of all geometrical objects: space itself. A Euclidean proof always begins with "Let there be a line," or a triangle, or a circle, or some such object. Then, as the proof proceeds, more lines, triangles, and circles are posited, and certain relations between them are established. But apart from these objects and the relations between them, there is nothing—not even emptiness. In this world, geometrical objects do not float in preexisting space but exist in and of themselves, and in relation to other objects when necessary. In as much as "space" exists at all in the Euclidean universe, it is simply the spatial relationship that the lines, triangles, and circles in a proof have to one another. The idea of an absolute space that exists in itself, before any geometrical figure is placed within it, was nonsense to classical geometers.

And the same is true for Descartes. Today we might think of analytic geometry in terms of an absolute geometrical space, defined by a set of coordinates. As many of us learned in high school, this universe can subsequently be filled with various geometrical shapes and curves, determined by their equations. Space itself, however, precedes any of them and can exist even when completely empty. But in the world of *The Geometry* there is no absolute space. As in classical geometry, objects exist in and of themselves and in relation to one another but not to "space" itself. Cartesian space, just like Euclidean space, is relative.

To understand the concept of relative space, consider, for example, sitting for dinner on a train racing through the countryside. Inside the train, things are more or less fixed in space: the table, the plates, the silverware, the seats, and the diners move very little within the interior space of the dining car. But a farmer standing in a field where the railway tracks stretch to the horizon would perceive things very differently: in the open space of the field, he himself is stationary, but the dining car and everything in it—table, plates, silverware, seats, and diners—are hurtling along at breakneck speed. Meanwhile, astronauts on their way to Mars, gazing back at the Earth, would perceive not only the train but also the farmer and the field to be moving, spinning rapidly around the Earth's axis while racing in orbit around the sun. For the diners on the train, all that matters is the space of the car's interior; for the farmer, all that matters is the space of the open field; and for the astronauts, all that matters is the space of the solar system. All of them are right in their own terms because each of them refers to a different relative space, a space defined by the relationships of the objects within it.

Relative space is the space of our daily lives, and it is therefore not surprising to find that it was the only kind of space considered by most ancient philosophers.[34] For most of those philosophers, as for Aristotle, the cosmos did not float in a vast void but existed in and of itself. The space within it was defined by the relationship of objects to one another, much like the space within the dining car. Beyond the cosmos there was nothing, no space, not even emptiness. Space was relative, and it stands to reason that if this is the case for the physical world, it would remain so for the abstract ethereal world of geometrical objects.

Now Descartes's world differed in critical ways from the ancient cosmos: whereas the cosmos was closed, the Cartesian universe extended in every direction; whereas in the cosmos different locations possessed different qualities, the Cartesian universe was homogenous, and all locations the same; and very much unlike the ancient cosmos, Descartes's world was mathematical through and through. But if in all these respects the Cartesian world seems like our own universe, in one respect it was profoundly different: for Descartes, spatial extension was indistinguishable from matter, meaning that all space was, by definition, full. In such a world, space simply cannot exist for itself independent of the objects within it; space exists by virtue of the objects that occupy it and is inconceivable without them. And so Cartesian space, like Euclidean space, cannot acknowledge a space that exists for itself, without objects occupying it. It is therefore, inevitably, relative space, defined by the geometrical objects that compose it and the relationships between them. And what is true of physical space must also be true of Cartesian mathematical space, as presented in *The Geometry*. Since the purpose of Descartes's reform of geometry was to describe physical space, it was inevitable that Cartesian mathematical space would be relative, just like its physical counterpart.

Two Straight Lines

The fact that Descartes's mathematics takes place in relative—rather than absolute—space shaped his analytic geometry in ways that sharply distinguish it from modern practice. Simply put, a geometry of relative space cannot make use of a fixed coordinate system, which is the defining feature of modern analytic geometry. There are, it should be noted, different kinds of coordinate systems, including polar coordinates and oblique ones, and all can be used in their appropriate context. But the simplest, the most common, and, in truth, the iconic set of coordinates is made up of two or—in the case of three-dimensional geometry—three perpendicular lines, stretching to infinity in each direction and intersecting at a single point. That they are known as "Cartesian coordinates" is an appropriate tribute to

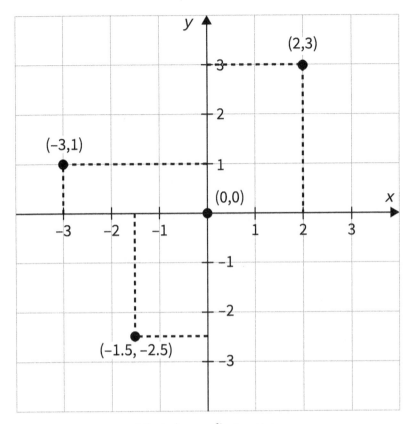

A Cartesian coordinate system.

the founder of analytic geometry. But it is also ironic, since Descartes never did, or would, use them.

To understand the significance of these coordinates, let us look at a simple diagram, familiar to many high school students today:

The horizontal x axis and the vertical y axis intersect at a right angle at a point known as the origin. This is the point where the values of x and y are both 0, and it is marked accordingly (0, 0). From the origin, the sequence of numbers increases uniformly to the right along the x axis and decreases to the left, whereas along the y axis, the numbers increase upward, above the origin, and decrease downward, below it. Accordingly, the integers are marked 1, 2, 3, 4 . . . at regular intervals to the right of the origin along the x axis and above the origin on the y axis. Similarly, the descending count of negative integers is marked $-1, -2, -3, -4\ldots$ to the left of the origin along the x axis and below it on the y axis.

This arrangement of two intersecting lines is almost laughably simple, and yet its power is remarkable: every point on the plane is now uniquely defined by these two coordinates. Take the point in the upper right of the figure above, for example: if we drop a line from it to the x axis, it will intersect with it at a right angle at the point marked "2"; and if we draw a line from the point to the y axis, it will intersect with it at a right angle at the point marked "3." Consequently, the point is defined as $(2, 3)$ in this coordinate system. The same is true of the points marked $(-3, 1)$ and $(-1.5, -2.5)$ and for every other point within the coordinates' plane. Every point is uniquely defined by a pair of numbers, each referring to a particular position along the coordinates.

The coordinates create a plane that stretches to infinity in every direction, a plane that is, in effect, an absolute space. Every point within it is defined not in relation to other points but in absolute terms, as the intersection at a right angle of a pair of lines stretching from specific points along the axes x and y. The same, furthermore, is true of any geometrical objects in the plane, whether a line, a triangle, a parabola, or any other figure. Each shape and each location are defined absolutely, in terms of its (x, y) coordinates.

It is the coordinates that establish this absolute space, and they do so by creating a rectilinear grid. Straight lines emanating from one of the axes at right angles and regular intervals intersect with those emanating from the other at identical intervals, forming a checkerboard pattern that in principle covers the entire infinite plane. The complete grid, seen in the figure above, is commonplace in elementary textbooks but is rarely marked in more advanced mathematical texts, which usually settle for displaying only the x and y axes. Advanced students, not to mention practicing mathematicians, do not require that every line of the grid and every square in the checkerboard be explicitly outlined. It is always implicitly there, however, effectively extending the coordinates and uniquely determining every point, shape, and location. This boring grid, monotonous, predictable, and stretching boundlessly in every direction, has become the very embodiment of mathematical space. It is infinite, uniform, and absolute, a perfectly regular space that exists in itself, everywhere and always, even when it is completely empty of any mathematical objects.

As is often the case with scientific or mathematical innovations, it is almost impossible to determine who "invented" our modern Cartesian coordinates. The use of two axes at right angles to each other seems to us so natural that it is hard to think how we could possibly proceed in mathematics without it. It has become, to us, self-evident, so much so that it hardly needs an "inventor" at all. It is simply common sense, the obvious and "natural" way of doing mathematics.

That, however, is not at all how things seemed to seventeenth-century mathematicians. Geometers had practiced their craft for two millennia without ever feeling a need for Cartesian coordinates or anything like them. Descartes applied algebra to geometry, but for him x and y, which we think of as coordinates, were simple lengths of lines and distances between given figures. Much the same was true of other mathematicians who contributed to analytic geometry, including his contemporaries and immediate successors: fellow Frenchmen Pierre de Fermat (1601–65) and Philippe de Lahire (1640–1718), the Dutch mathematicians Frans van Schooten (1615–60) and Jan de Witt (1625–72), and the Englishman John Wallis (1616–1703).[35] None of them posited an absolute mathematical space through the use of fixed rectilinear coordinates, the scheme that we find so obvious and natural. The reason is simple: Mathematicians of that era thought of geometrical objects as residing in a relative space, defined by the objects themselves and others in their immediate vicinity. And as long as that was the case, there was simply no reason and no place for the universal coordinates we take for granted.

It is therefore no coincidence that the first mathematician to systematically make use of Cartesian coordinates was also the one who also advocated fiercely for the reality of absolute physical space: he was, of course, Isaac Newton. Absolute space, as we have seen, was for Newton the fixed, uniform, and infinite space that pervades the entire universe and makes room for God to act in the world in ways that were impossible in Descartes's "full" universe. And it was this absolute empty space that both requires and allows for the workings of the universal and mysterious force of gravity, which Newton believed originated in God's will. Absolute physical space was key to Newton's theory of the physical world; and since this world was mathematical through and through, it should come as no surprise that Newton's mathematics was based on the same assumption: that geometrical objects exist in an absolute mathematical space—uniform, universal, and infinite.

Charting the Void

A careful reader might reasonably interject here: It is all well and good to point to Newton's novel concept of "absolute space" and the key role it played in his science, philosophy, and even religion. But does it really follow from this that Newton's *mathematical* space would have similar characteristics? Modern mathematicians, after all, have no trouble distinguishing between mathematical space, whose characteristics might vary in accordance with our assumptions, and the actual physical space we live in. Wouldn't the same be true for Newton?

The answer, however, is no. Far from distinguishing between mathematics and the physical world in the manner of modern mathematicians, Newton insisted that they are one and the same. Geometry, he wrote in the preface to the *Principia*, is founded on the physical mechanical practice of drawing circles and lines and cannot be separated from it. It is nothing, he argues, "but that part of universal mechanics which accurately proposes and demonstrates the art of measuring."[36] For Newton, mathematical objects are actual objects in the real world, and mathematical relations are actual relations in the real, physical world. The absolute space of the physical world is also, accordingly, the absolute space of the mathematical one.[37]

That for Newton geometrical curves and figures exist in an eternal absolute mathematical space is eminently clear in his works on analytic geometry. The most important of these is a treatise titled

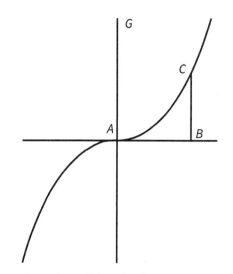

Newton's graphing of the cubic function in *Enumeratio*.
Newton, *Enumeratio*, fig. 77.

Enumeratio Linearum Tertii Ordinis (Enumeration of lines of the third order), first published in 1704, though based on materials from decades earlier.[38] In the *Enumeratio*, Newton sets himself the goal of classifying all forms of cubic curves (or "functions," as we would call them today), which include not just the familiar $y = ax^3 + bx^2 + cx + d$, which we commonly call the "cubic function," but all variations that may also include ey^3, fy^2x, gyx^2, and so on. All in all, Newton identifies seventy-two different classes of these of the equations, and he draws the graphs of each and every one. Some of these are simply bizarre, including separate branches shaped as parabolas, hyperbolas, and ovals. The seventy-second and last category is the common cubic function, for which Newton graphs the simplest form, $y = ax^3$.

This graph, like all other graphs in the treatise, is structured around a pair of open-ended rectilinear axes that implicitly extend to infinity. In fact, the *Enumeratio* is probably the first work in the history of mathematics to systematically construct and utilize the two "Cartesian" axes.[39] These axes are not dependent on the specific curve, as is the standard practice in Euclidean geometry and in Descartes's constructions, but exist in themselves. The coordinates are there regardless of which curve is drawn on them or even if any curve is drawn at all. Even in complete emptiness they will still be there, establishing an infinite, uniform, and absolute space in which geometrical figures can be drawn. In the *Principia*, Newton established the concept of absolute physical space, which exists in and of itself, before any stars, planets, or connecting forces are placed within it. In the *Enumeratio*, he does the same for geometry, establishing an absolute geometrical space that exists before any circles, ellipses, parabolas, or other shapes and curves are placed within it.

Why does this matter? Because, as Newton and his followers powerfully argued, absolute empty space allows for something that relative and full space does not: freedom of action within the universe. For Descartes, it will be recalled, matter was identical with extension in space, and empty space was therefore impossible. In his world material objects rubbed up against other material objects, pushing and pulling them to create the perfect mechanism that is our world. For Newton, however, such a picture was unacceptable because it did not leave room for God to intervene in the world. Only an absolute preexisting space, according to Newton, gives full scope for God's action, allowing him to create the stars and planets as he sees fit and establish the laws and forces that connect them, and in particular gravity.

And what was true of God's freedom was also true for man, as we learn from one of Newton's most celebrated admirers, John Locke (1632–1704). After positively reviewing the *Principia* in 1688, Locke met and befriended Newton the following year and was so impressed with him that he began referring to him as the "incomparable Mr. Newton," both in his correspondence and in the preface to the *Essay Concerning Human Understanding* of 1690.[40] It is in that treatise that he famously proposed that the human mind is a "blank slate," or, as he put it, a "white paper, void of all characters, without any *ideas*," which is gradually inscribed by experience.[41] Locke initially argued, in good Cartesian fashion, that the mind is affected by material impulse, or direct contact, because this is the only way in which matter can act on matter.[42] But when warned by Bishop Edward Stillingfleet that a mechanical world ruled only by impulse leaves no room for either

God or man to exercise free will, Locke reconsidered. "So I thought when I writ it," he explained, but his views were subsequently transformed by Newton's account of gravitation, acting through the void. For Locke, it is precisely this unconstrained emptiness that allows both God and man to be free.[43]

And just as absolute physical space allows freedom in the physical world, absolute mathematical space does the same in the mathematical world. Euclidean geometry was powerful and elegant but also limited, allowing only figures that could be constructed by compass and straightedge. Consequently, the Euclidean world was a completed world, with no new objects added since it was first systematically described in the *Elements*. Descartes's geometry was more flexible, but it too relied on carefully positing certain lines and locations and then—using algebra—describing the relations between them. But it was Newton's geometry of absolute space that finally broke through the traditional strictures. As the *Enumeratio* shows, every algebraic equation defines a graph, and the range of shapes produced by such formulas is practically unlimited. God created the sun, moon, Earth, and planets in the vast emptiness that is absolute physical space. The geometer can similarly create not only the traditional conic sections but also things never before dreamed of, such as cubic and semicubic parabolas, and hyperbolas with "cross-shaped," "conchoidal," and "snaking" branches.[44] By simply posing an algebraic equation, a geometer can play God in absolute mathematical space, creating new and unheard-of objects out of nothing.

Absolute geometrical space, in other words, grants the geometer the power and freedom to establish a new geometrical world, inhabited by novel geometrical objects. Yet powerful as it is, establishing this geometrical space is easy: all it takes is positing two rectilinear axes to mark an absolute two-dimensional plane, or three axes to mark an absolute three-dimensional space. As one can see in any of Newton's diagrams in the *Enumeratio*, the fixed axes divide the infinite plane into four rectilinear "quarters." Each of these, furthermore, is implicitly covered by a rectilinear grid of lines parallel to the two axes. Wherever two such lines meet in the plane, they define a unique point, absolutely, regardless of whether the point is occupied by a curve or by nothing at all. It is this grid, sometimes depicted fully, sometimes only in outline, that is the hallmark of absolute geometrical space. It is this grid, in other words, that opens up the space of freedom.

Newton's notion of geometrical space was radical in its time, but it did not remain so for long. In the decades that followed, the absolute rectilinear space of geometry quickly became for mathematicians what it is for us: natural and self-evident, so much so that it hardly needs remarking on. We see it in textbooks, such as Colin Maclaurin's *Treatise of Algebra* of 1748, which teaches students how to construct curves in the absolute space defined by rectilinear axes, providing example after example in a manner reminiscent of the *Enumeratio*. We see it also in the most advanced and sophisticated work of the greatest mathematicians of the age, such as Leonhard Euler. In his classic *Introductio in Analysin Infinitorum* of 1746, for example, he explains how three planes intersecting at right angles "divide universal space into eight parts."[45]

By the middle of the eighteenth century, absolute Newtonian space had become the natural domain for mathematics, the place where geometers traced lines and curves and where geometrical objects had their ethereal—yet very real—existence. It extended to infinity in all directions; it was uniform and homogenous, everywhere the same; and it was absolute, meaning that every point in space was precisely and uniquely defined. Wherever it appeared, it was easily identifiable by the permanent rectilinear axes and the grid—implicit or explicit—that covered it from end to end. It was this grid that for geometers and students alike was the mark of empty mathematical space, the space in which they would produce their mathematical creations.

The rigorous inner structure of Euclidean geometry made it a potent weapon in the cultural and political battles of the early modern world. Perfectly rational, rigorous, hierarchical, and unchallengeable, the world that Euclid made was a powerful representation of the kind of state rulers such as Louis XIV sought to create. At Versailles, the Sun King built a Euclidean world that demonstrated that absolute royal supremacy was as natural, as rational, and as unchallengeable

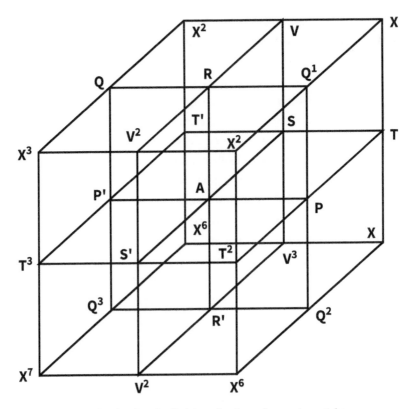

Leonhard Euler, the division of universal space into eight
rectilinear parts, *Introductio in Analysin Infinitorum. Introductio in
nalysin Infinitorum*, vol. 2, fig. 120.

as a geometrical proof.[46] But what of the world presented by the Newtonian geometry of empty absolute space? For a king such as Louis
XIV, such geometry would not do. For him, the power of Euclidean
geometry was precisely that it created a single irrefutable world order,
which he ingeniously associated with his own rule. The new rectilinear geometry, however, did nothing of the sort: instead of the rich Euclidean world in which every object had its precisely assigned place, it
offered only empty space; and instead of the unique unavoidable and
unchallengeable order, it offered infinite possibility and the freedom
to utilize it. For unlike the figures of Euclidean geometry, the familiar rectilinear grid did not suggest a complete and irrefutable world
order: it was nothing but a blank slate on which geometers could cre-

ate their figures and write their stories. Euclidean geometry stood for a single, necessary, irrefutable, and unchanging order, just as Louis would have wished; the Newtonian geometry of empty space stood for the possibility and opportunity to create a new world however one wished. The very idea of such a world would have appalled the Sun King and sent shivers through the spines of the great princes of Europe.

Yet while Bourbon France remained wedded to its Euclidean vision, there was even then a land where the implications of the new geometry were appreciated and welcomed. It was a land so large it appeared endless, if not infinite, to its European inhabitants, a land that was as empty in their eyes as Newton's absolute space, even though it had been, in truth, settled for a thousand generations. It was a land devoid of Europe's history, its traditions, and the inescapable hierarchies and customs that define that continent's societies. It was a land of possibility and opportunity, a blank slate on which new men would live freely and create a world different from any that had come before. It was, in other words, a land that was ideally suited and ready to embrace the promise of the new geometry of empty space. It was, of course, America.

2

....

Life, Liberty, and Infinite Space

The Eye in the Sky

Try a simple experiment on your computer. Open up Google Maps, and in the location bar enter "Iowa, USA." An outline map of the state will emerge in the center of your screen, roughly rectangular in shape, bounded by the Mississippi River to the east and the Missouri River to the west, with straight-line boundaries to the north and south. Only two main roads are initially marked in this map—Interstate 35, running north to south, and Interstate 80, running east to west. They intersect at a right angle in the capital, Des Moines, about two-thirds of the way down the I-35. Zoom in a bit more, and more details will reveal themselves: two more east–west routes, US 20, which intersects with Interstate 35 halfway between Des Moines and the northern border with Minnesota, and US 30, running through the middle of the state and intersecting Interstate 35 at its geographical center in the city of Ames. Smaller roads, parallel to the main axes and at more or less regular intervals, also appear. Together they carve up the larger rectangle of the state into a rough pattern of smaller rectangles. These, a quick check will confirm, correspond almost precisely to the ninety-nine rectangular counties that make up the state of Iowa.

As we continue to zoom in, the pattern repeats itself over and over, even more regularly and more precisely. Switching to the satellite imagery of Google Earth, we now detect many more roads, all of them parallel to the main axes, running either north–south or east–west, and at precisely regular distances from one another. At this scale, it becomes clear that these are not just roads, built to bring people from place to another, but also boundaries that separate one plot of land from its neighbor. Perfectly straight and regularly spaced, these boundary roads create between them a checkerboard pattern of square fields, mile after country mile of one square field following another in endless succession from one horizon to its opposite in every direction. Looking down from Google Earth's eye in the sky, we are struck by the stunning regularity and monotony of a pattern that transforms a natural landscape into something else entirely. In driving through the country roads of Iowa, one has the experience of physically inhabiting an abstract mathematical grid, as precisely measured and outlined as a sheet of graph paper.

And just as a mathematical grid continues to infinity, uncontained by border or boundary, much the same is true for the midwestern grid of roads and fields. As seen from Google Earth, the political and administrative boundaries of the state of Iowa disappear altogether, leaving only the spread of the grid itself, which repeats itself endlessly in each direction. It spreads northward to Minnesota, southward to Missouri, eastward to Wisconsin and Illinois, and westward to South Dakota and Nebraska. All in all, this Great American Grid covers no less than two-thirds of the continental United States, stretching from Ohio in the east to California in the west.

It is an area of staggering size. It contains great rivers—the Ohio and Tennessee in the east, Mississippi and Missouri in the center, Columbia and Colorado in the west—and great mountain ranges, from the slopes of the Appalachians in the east to the Rockies and Sierra Nevada in the west. It covers regions with distinctly different cultural, ethnic, and political histories: from Minnesota and the Dakotas, settled by German and Scandinavian immigrants, to states of the Old South, such as Mississippi and Arkansas, where the wounds of slavery and the Civil War have never fully healed. It stretches over the Southwest, where New Mexico and Arizona cling to their unique Hispanic and Mexican heritage, and on to the Pacific coast, where the most populous state, California, is perpetually struggling to forge an identity out of a fabulously diverse citizenry. And everywhere it covers what was once open native land, whose people have now been confined into dense and strictly bounded reservations.

Here and there, the great grid is forced to make concessions in the face of both natural and human opposition. Great rivers are natural boundaries that cannot be ignored, even though they never flow in straight lines and rarely align with the points of the compass. And so Iowa, bounded by the Mississippi and the Missouri rivers, is not a perfect square like Colorado and Wyoming but only a rough approximation of one. Mountain ranges such as the Rockies, furthermore, simply cannot be carved into regular squares, and so, wherever the terrain is too rough and ragged, it is inevitably left undisturbed. Local politics and culture also make some inroads. In the northern and western territories, the surveyors charged with creating the grid were on the ground with their instruments before most settlers arrived, and were therefore free to create the perfect grid that stretches across the Great Plains. But in the South, the same surveyors, armed with their government writs and licenses, confronted established polyglot communities of French and Spanish, as well as Native Americans and English settlers. The resulting grid, showing the marks of this struggle, is not nearly as regular and precise as it is farther to the north.[1]

Most of the intrusions into the great grid pattern, however, were small and local: a creek here inconveniently intrudes into a precise square, a lake there joins two squares together, a jagged rock formation cannot abide the rigid regularity of the grid, and so on. Then there are the towns, inevitable even in a rural setting, for which the uncompromising grid makes no accommodation. Unlike European towns and cities, which are the centers of star-shaped convergences of roads that cut through the countryside, most western American towns bow to the demands of the imperious grid. Their main avenues are extensions of the country roads that divide one square plot from another; their rectilinear and evenly spaced streets run parallel to the grid lines, toward the four points of the compass. As they spread out from their centers, American towns tend to blend seamlessly back into the Great Western Grid, causing barely a ripple in the endless repetition of its monotonous pattern.

Because all in all, there is no escaping the great grid that covers the western two-thirds of the United States. It may disappear for a stretch, at the banks of a river or the foot of a great mountain range. It may fade away temporarily at the boundary of great cities such as Indianapolis or St. Louis, metropolises that adhere to their own urban plans. But it reappears again: across the river, in confined valleys sheltered among towering peaks, where its uniform squares briefly appear before dissolving into the surrounding mountain slopes. Beyond the mountain ranges and past the great cities, it reappears as if it had never vanished in the first place. The same exact lines that had faded out will reassert themselves miles away across rivers, mountains, and cities, arrow straight and oriented to the points of the compass, dividing the land into a single uniform pattern of squares that has no beginning, no middle, and no end.

The Land before the Grid

It wasn't always so. For the American West had been home to the native peoples of America for countless generations before the first straight line was surveyed and the first right angle was marked on the land. And to them, the great grid was not just alien and forbidding, an unwelcome interloper in the country of their ancestors; it was all but incomprehensible. Their own conception of the land they lived in and their place within it made nonsense of the grid's arbitrary division of the land into neat rectilinear parcels. To Native Americans, each and every place had a unique significance that could not begin to be captured by the intersection of two coordinates.

Native Americans, it hardly needs saying, were not and are not a single people. Among them are many hundreds of different tribes and nations, speaking different languages, following different customs, and possessing different religious outlooks. The Algonquian farmers of the Atlantic Seaboard, the nomadic Sioux who followed the buffalo herds on the Great Plains, and the Chumash fishermen of the California coast have far less in common than the various European peoples who so pride themselves on their national peculiarities. Making generalizations about Native Americans is therefore inherently hazardous, for exceptions can almost always be found.

Yet for all these differences, this much is true: the native peoples of America did not view their land as abstract space, in which every location was interchangeable with any other. Quite the opposite: every hill in their territory had its own history, going back to the time of their ancestors; every path had its own stories associated with it, its own shared memories, myths, and moral lessons; every cave and forest grove was home to its own spirits, benevolent or malign, and thereby its own powers to shape human life. The land was not a space within which people's lives, traditions, and social relations unfolded. It was, rather, a living part of these relations and inseparable from them. The land is imbued with social, historical, moral, and religious meaning, and it is an active participant in the life of its people. It is the polar opposite of the abstract, passive, empty space that Newton imagined for his universe. It is the polar opposite of the monotonous and uniformly gridded West.[2]

An indigenous map given in 1721 to Francis Nicholson, governor of South Carolina, echoes across the gulf separating these two incompatible notions of the land.[3] Etched on a deerskin, it was the gift of the inhabitants of a group of Indian villages not far from the capital of Charleston, who later became known as the Catawba. The only reason it survives today is that Nicholson had a paper copy made and forwarded it to his superiors in London. As far as modern scholars can tell, maps of this kind were practically unheard of within native societies. They were not used for transmitting knowledge among members of close-knit groups or for communicating more broadly between different tribes or nations. The fact that the Catawba produced one at all is a measure of the effort they made to accommodate what they understood to be a European perspective.[4] When they gifted it to Nicholson, the leader of the white colonists in their region, they were making a mighty effort to reach across the cultural divide and communicate in a visual language that they hoped the Englishmen would understand.[5]

A quick glance at the deerskin map suggests that in many ways the Catawba were impressively successful. For indeed there is much here that corresponds to European notions of a "map." Just like a European chart, the native map is intended to represent a specific geographical region in an abstract manner, indicating its most important features. There are circles marking the eleven Catawba villages, double lines indicating the pathways between them, and regions marked "Virginia" and "Charlestown."[6] The parachute-like object at the bottom left is, in fact, a ship with a mast and stays, indicating the presence of a harbor. All this is as one would expect in any European-style map, and it seems likely that the Catawba cartographers had seen such objects before setting out to produce one of their own. And yet despite their best efforts, the map they produced is far from being the monument to intercultural communication they likely hoped for. If anything, it points to a deep chasm between two outlooks that is all but unbridgeable.

For the deerskin map, at every level, is structured by an outlook of the land and its people that is all but incompatible with the European view. First there are the Catawba villages themselves, represented by a set of circles. Each circle was not merely a geographical location but a "fire," referring to a community bound by blood and custom.[7] To learn about a fire, it was not enough to look at a map—one had to know its lineage, its traditions, and its relationship to other fires. The fires, in turn, differ in size and are connected to one another by pathways, which to a European-trained eye appear very much like roads on which one can travel from one village to another. That, however, would be a mistake, for the pathways marked on the Catawba map are of social and political—not physical—connections. The fire called Nasaw, for example, is depicted as the largest and most important of the Catawba settlements because of the size of its circle, the centrality of its position, and the number of its connections to neighboring settlements. Other fires reside on the periphery of Catawba society

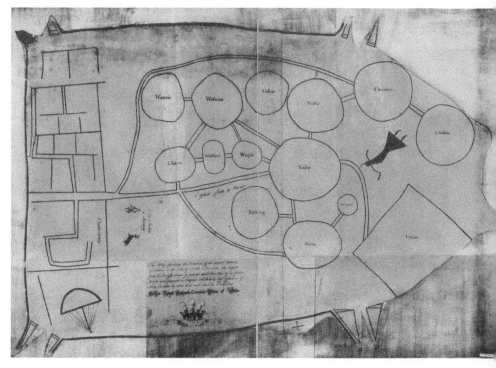

The Catawba deerskin map of 1721. Courtesy of the Library of Congress,
Geography and Map Division.

and have only one or two connections; Nasaw has seven, making it
the gateway to the entire network of villages.

And just as Nasaw is at the center of the Catawba fires, the Catawba
villages as a whole are at the center of a wider network of peoples. The
Cherokee and the Chickasaw are depicted as their own fires on the
margins of the Catawba polity, connected to it by a single and (in the
case of the Chickasaw) indirect pathway. The European settlements
are, if anything, even more marginalized: a rectangle marked "Vir-
ginie" occupies the bottom right corner and is connected to Nasaw by
a single pathway. Charleston, despite being a city, is depicted as larger
than Virginia and occupies much of the left side of the map. And like
Virginia, the only other European entity depicted, it is marked not by
fire circles but by intersecting rectilinear lines, likely inspired by its
gridded street plan.[8]

Taken together, it becomes clear that the Catawba map is designed to convey something very different from a European map of the same territory. A traveler trying to make their way from the village of Waterie, for example, to Charleston using the native chart as a guide would quickly become lost. Not only does the map lack roads and natural landmarks but it doesn't even have fixed directions! The various orientations of the human and animal figures, as well as the inverted ship at the bottom left, clearly indicate that the map is intended to be looked at from all sides. Directions are thereby not absolute but relative to one's position and point of view. Yet though hopelessly disoriented, the traveler would nevertheless learn other important things: that the villages of the Catawba are the most important settlements in the region; that certain villages are related to certain other villages but not to all; that Nasaw is the most central and connected village of all; that some of the Catawba's neighbors (Cherokee, Chickasaw) are in certain ways like them and are accordingly marked by fire circles; that others (Virginia, Charleston) were not and were therefore marked differently, with straight lines and right angles. All of it is information that the traveler would find highly useful when traveling in the land of the Catawba.

To some extent, one could argue that a European map focuses on physical features of the land whereas the native depicts social relations. But this would be misleading. The Catawba villages are, after all, physically real and really do occupy a region bordering the lands of the Cherokee and the Chickasaw, and the English colonies of Virginia and Charleston really are nearby. The rectilinear streets of Charleston are also physically there, as is the town's harbor. The Catawba map, in other words, just like a European one, does depict a real physical landscape. It is just that for the native people who produced the map, this landscape is inseparable from the humans who inhabit it, their traditions, and their relations. The land defines human connections, placing the Catawba at the center of their world and Nasaw at the center of their villages; conversely, human practices and histories give meaning to different locations, as the various fires with their interconnected pathways and the rectilinear streets with their borders imprint themselves on the land.[9]

For the Catawba and other native peoples, humans were shaped by the land, just as the land was marked by its people. One could not be without the other. The notion of a land that is empty, uniform, and devoid of human meaning, and of a people devoid of roots and history in the land they occupy, was to them nonsensical. And yet the Great Western Grid proclaimed precisely that, imprinting a cold and impersonal mathematical pattern on a staggering 1.4 billion acres of the North American continent. The tribes resisted as much as they could. In 1851, for example, the Lakota rejected the demand to grid their territory on the grounds that it amounted to being told "to suddenly stop being what they were," and to a similar demand made in 1877 to the Nez Percé Chief Joseph, his spiritual adviser Toolhulhulsote responded that "the earth is part of my body, and I never gave up the earth." But in the end they proved helpless to stop the juggernaut of the grid as it rolled across the West, transforming a living landscape into a blank mathematical space. The consequences for the native people of America were, as we shall see, catastrophic.[10]

A Land like No Other

It is hard for Americans, and especially westerners who have lived all their lives on the surface of the endless mathematical grid, to appreciate just how strange and unlikely a landscape it is. If we compare it with any European agricultural landscape, we see immediately that there is nothing like it in the Old World. The French countryside, viewed from above, shows a beautiful but irregular pattern of fields and roads, accommodating natural features, no doubt, but mostly historical circumstances. For the fields, in their day, were parts of noble estates and clustered around castles and manor houses, most long gone but some still present. In as much as a pattern can be detected at all, it is of roads converging diagonally through the countryside toward market towns or former manors, where they meet in star-shaped intersections. Though hardly formal or elegant, these interlocking stars yet faintly echo Le Nôtre's elegant geometrical patterns at Versailles and may have served as their inspiration.

A somewhat more familiar pattern can be found in Italy's Po Valley, where the boundaries of fields descend from a rectilinear pattern installed by the Romans for their military colonies. Straight lines and right angles can also be divined in the plots of the Dutch lowlands and in the Japanese Jori system. Yet even those supposed "grids," left to us by the ancient masters of management and organization, are limited in extent and variously oriented in accordance with the peculiarities of the local landscape and pattern of settlement. Even if one looks beyond the accumulated irregularities of two millennia, it is clear that there is nothing here of the precision, regularity, and uniformity of the American grid, not to mention its limitless scale and ambition to overlay an entire continent.[11]

In its simplicity, uniformity, and familiarity, the Great American Grid appears to us entirely self-evident, the most rational and simple way to organize our world. And yet, when we compare it to pretty much any other place, we realize that there is nothing natural or self-evident about it. It is grand, ambitious, and uniquely American, and though rarely noted or discussed, it is nevertheless one of those inimitable constructs that sharply distinguish the United States from the ancient European monarchies that spawned it.[12]

When people speak of "American exceptionalism," some refer admiringly to the United States as an egalitarian nation of immigrants that unlike other nations is not founded on a shared history or ethnic bonds but on shared democratic ideals expressed in the Declaration of Independence and the Constitution. Others speak disparagingly of American arrogance and Americans' righteous conviction that they are always on the side of the good and right, no matter how violent the methods they employ. People hardly ever speak of the fact that unlike its European forebears or practically any country in the world, America is a gridded land, carved up from top to bottom into a single massive and regular pattern of evenly spaced rectilinear lines, running north to south and east to west.[13]

And if it would seem that this Great Grid is merely a superficial overlay on the land, a bureaucratic convenience that tells us little about the country and its people, consider this: the author of the Great

American Grid is also the author of the Declaration of Independence. The man who insisted on carving an entire continent into tiny regular squares is also the man who did more than anyone to formulate and express Americans' highest ideals and shape his countrymen's view of themselves. He is the one whose eloquent tributes to American freedom have lost nothing of their power to stir the heart yet one who was also a slaveholder, who fathered children with his enslaved mistress and did not hesitate to sell or break up families when he needed to cover his debts. He is the man who both admirers and detractors of the United States point to as the emblem of the American experiment, the most beloved of the Founding Fathers after Washington himself and the most reviled: Thomas Jefferson.

The Beckoning Lands

Looking back from our twenty-first-century vantage point, it is hard for us to conceive of a connection between the systematic carving up of American land and the political ideals for which Jefferson is known to this day. But for Jefferson himself, who was surrounded by the maps and tools of land surveying from the day he was born, the two were inseparable. Born in 1743 in Shadwell, Virginia, the future president was the son of Peter Jefferson (1708–57), one of the most accomplished land surveyors and cartographers in all the American colonies. In 1746, Peter Jefferson was employed by Lord Fairfax to survey his five-million-acre Virginia estate, which stretched from the lowlands to the Allegheny Mountains. Some years later he collaborated with fellow Virginian Joshua Fry (1699–1754) on the most accurate map to date of the colonies of Virginia and Maryland, which was published in 1752 and became known as the "Fry-Jefferson Map." Toward the end of his life, the elder Jefferson moved from being a professional surveyor in the service of others to being an active land speculator in his own right. Along with his neighbors in Goochland County, Virginia, he organized the Loyal Land Company and set about surveying the territory west of the Alleghenies with a view to selling plots to land-hungry settlers. It was a bold venture, but a sound one too, since Jefferson Sr. was capitalizing on a fever for western lands that had seized the colonies and would not abate for another century and a half.[14]

Peter Jefferson and his associates were far from alone in attempting to capitalize on western land, which proved an irresistible lure to settlers not only from the colonies but from as far away as England and continental Europe. Others who enthusiastically took part in this lucrative business included legendary frontiersman Daniel Boone, the future Founding Father Patrick Henry, and the commanding colonel of the Virginia militia, one George Washington. It is sobering to remember that in the 1750s and 1760s, the future hero of the revolution and saintly first president of the United States was making his fortune in risky land deals of dubious legality.[15]

As Peter Jefferson, George Washington, and their fellow land speculators saw it, America—with the exception of the colonies themselves—was an empty land without a people. Once one crossed the great ridge of the Appalachian Mountains, which formed the western boundary of the settled colonies, one entered a no-man's-land that stretched all the way to the Pacific Ocean. In this great void, they reasoned, an enterprising man had every right to survey the land, enclose it against all comers, declare it his property, and buy and sell it as he saw fit.[16]

This was, undoubtedly, a self-interested conceit on the part of men set on making their fortune by selling vast tracts of land they had rarely, if ever, seen. And it came up repeatedly against the stubborn reality that land was not, in truth, empty but had been settled by native people for countless generations. And yet the surveyors and settlers believed themselves to be speaking the truth when they declared the continent empty and its land awaiting its European settlers. They were, after all, continuing in a long American tradition. Even John Winthrop, first governor of the Massachusetts Bay Colony, had observed before setting off to America that since the Indians did not divide up the land with clear boundaries and keep out interlopers, they did not have any claim to it. "As for the Natives in New England," he wrote, "they inclose noe land . . . so as if we leave them sufficient for their use, we may lawfully take the rest."[17] As far as the surveyors, speculators, and settlers were concerned, the Indians were an annoyance, an irritant to be swept aside. As regards land ownership, their rights would be contested, ignored, and ultimately denied.[18]

Yet for all that, the Indians refused to go away. The territory west of the Appalachian Mountains was nominally French, a fact that was recognized in international treaties but had only limited impact on the ground. This is because the entire French population in the Americas in the 1750s numbered in the tens of thousands, whereas that of the British colonies had long surpassed one million. And so when British surveyors such as Peter Jefferson and George Washington were scouring the territory with an eye to enclosing land and selling to British settlers, they encountered few Frenchmen but numerous Iroquois, Shawnee, Cherokee, and other native tribes.

When war between the English and the French came in 1754, the tribes were forced to choose sides. Some chose the French, whose rule was more nominal than real and allowed the tribes to pursue their way of life with little interference. They fought bravely and effectively and managed to hold off the British for several years before ultimately succumbing to overwhelming odds. Other tribes sided with the British, who with their huge population advantage seemed like a safer bet to emerge victorious. As members of the winning coalition, they reasoned, they would be better positioned to influence the terms of the settlement that would inevitably follow in the war's wake.

And so, indeed, it proved, initially at least. On February 10, 1763, Britain and France signed the Treaty of Paris, which among its other provisions ended French claims to North America. New France, which had stretched from the mouth of the St. Lawrence River in Canada to the Gulf of Mexico and westward to the Missouri River basin, had ceased to exist, and its territories east of the Mississippi had been transferred to Britain. To the residents of the thirteen British colonies, it was a historic victory, which finally delivered them from the threat of attack by French forces and the depredations of their Indian allies. With the territory beyond the Appalachian Mountains now securely in British hands, they naturally assumed that they were free to settle the land. It was the fruit of victory, as well as a just reward for the sacrifices they had undertaken on behalf of their king during the long years of struggle.

The king, however, had other concerns. Although he was now officially the master of vast new lands in North America, he was well aware that his actual hold on the territory was nominal at best. If British settlers were granted free access to western lands, the king and his ministers feared, they would soon move beyond the reach of both British officials and British commerce. Within a few years they might start producing their own goods, in competition with the mother country, or even set up independent polities. And if this were not enough, allowing settlers to pour into the Ohio Valley and Kentucky would betray the trust of the tribes who had fought shoulder to shoulder with British forces, not to mention permanently alienate those tribes who had fought alongside the French. It would not be long before the king was confronted with unruly settlers on the one hand and embittered native tribes on the other. His claim to rule the territory would turn into an empty boast.[19]

And so, on October 7, 1763, a mere months after signing the peace treaty in Paris, George III issued a proclamation that he believed was essential for preserving the fruits of the British victory in America: "It is just and reasonable," the king declared, "and essential to our Interest, and the Security of our Colonies, that the several Nations or Tribes of Indians with whom We are connected, and who live under our Protection, should not be molested or disturbed in the Possession of such Parts of Our Dominions and Territories as . . . are reserved to them." Consequently, he continued, we "declare it to be our Royal Will and Pleasure that no Governor or Commander in Chief in any of our Colonies . . . do presume, upon any Pretence whatever . . . to grant Warrants of Survey, or pass Patents for any Lands beyond the Heads or Sources of any of the Rivers which fall into the Atlantic Ocean from the West and North West, or upon any Lands whatever, which . . . are reserved to the said Indians, or any of them."

With the stroke of his pen, the king brought the expansionist dreams of his white American subjects to a quick and brutal end. He drew an impenetrable line along the highest peaks of the Appalachian Mountains, a line beyond which no surveyor, speculator, or settler might pass. The lands beyond, recently acquired from France, were reserved for the Indian tribes and would henceforth be protected from trespassing by the king's subjects. The royal government was no doubt well aware that men such as Peter Jefferson, Daniel Boone, and George Washington were not known for their respect for authority. In previous decades they had paid absolutely no heed to the dictates of French authorities and had conducted their surveys in the firm belief that the weight of numbers and of British arms would see them through. And so the royal decree made it as explicit as possible that times had now changed: no "pretence whatever," it makes clear, would be accepted as an excuse for surveying, enclosing, or registering the newly designated tribal lands. And as for British arms, so recently mobilized to protect the colonists, they would now be just as effectively mobilized to prevent the colonists from infringing on the rights of the Indians.

George III never doubted he had every right to quash the colonists' dream of western expansion if he saw it as in his interests to do so. America, for the king, was a territory like any other in his growing domains, a valuable asset in his ongoing competition with France for global supremacy. Like Jamaica, Barbados, or the subcontinent of India, it should be evaluated for its economic and political worth. Now that vast new American domains had passed into British control, it was obvious to the king that the Indians were as much his subjects as any of the many millions in his empire. The land was his to dispose, and if it suited his interests to bestow it on his new subjects, there was nothing more to be said on the matter.

The colonists, however, were in shock. America for them was not a colonial possession to be disposed of according to the political interests of the crown. It was their patrimony, the promise of their future. And whereas for the king the land was already occupied by its rightful owners, for the colonists it was a great and empty land awaiting its settlers. The notion that the Indian tribes roaming the vast interior of North America somehow held title to the land was absurd, the brainchild of a king and his council who had never set foot in the New World. As the colonists saw it, they had fought for the land, and now, at the moment of their triumph, they were being robbed of the fruits of their victory by the arbitrary proclamation of a distant monarch. It was an outrage, a great injustice that could not be allowed to stand.[20]

There were, no doubt, many roads that led from the moment when king and colonists stood together in triumph over their vanquished foes in 1763 to the violent rupture between them only thirteen years later. There was the clash over tariffs, the efforts by the London government to regulate the colonists' commerce, the fight over taxation, and even greater fight about representation. And there was the core question of whether the king, in ruling the colonies by royal decree, was depriving his American subjects of their rights as freeborn Englishmen. Yet surely one of the main issues that drove George III and the colonists apart, and ultimately into armed conflict, was the fact that they viewed America itself and their place within it in ways that were incompatible. The victory in the Seven Years' War and the addition of New France to the domains of the king of England had exposed a rift with the colonies that would not be healed. Not until, that is, the connection between the American colonies and the king in London was permanently severed.

In the meantime, the colonists did their best to ignore the royal edict, assuring themselves that it could hardly be enforced and would soon be repealed. The companies that were formed before the ban to survey and sell western lands continued to operate, and new ventures were formed. The Mississippi Company, which counted George Washington among its founders, was established in 1768 to settle the land along the great river, and the Illinois and Wabash Company, the Watauga Association, and the Vandalia Company followed suit. Each sought to survey, enclose, and sell millions of acres of western lands to eager settlers from the colonies.[21] It might seem surprising that wealthy merchants and landowners from the colonies were willing to risk vast sums of money on ventures that openly defied the royal edict. Since the ventures were blatantly illegal, the authorities could, in principle, shut them down at any moment, and their investors would lose everything. But the fact is that since the investors could never accept the official view that the seemingly boundless western lands were Indian territory, they convinced themselves that the ban could not possibly stand the test of time or survive the realities on the ground.

Washington's attitude was no doubt typical. In 1767, he wrote to a fellow Virginia planter encouraging him to join in "attempting to secure some of the most valuable Lands in the King's part, which I think may be accomplished after a while notwithstanding the Proclamation that restrains it at present and prohibits the Settling of them at all." The reason, he continued, is that "I can never look upon that Proclamation in any other light . . . than as a temporary expedient to quiet the Minds of the Indians." It would not last, he assured his friend, "but must fall of course in a few years," and it was therefore essential that they take action now. "Any person therefore who neglects the present opportunity," he warned, "will never regain it." To avoid complications, however, Washington asked that his friend "keep this whole matter a profound Secret . . . because I might be censurd for the opinion I have given in respect to the King's Proclamation."[22]

Yet despite the finest hopes of the colonists, who continued to pretend that the king's order was of no consequence, the ban stood. And as long as this was the case, the surveying and settlement of western lands was effectively illegal, land sales could not be registered, and anyone purchasing a plot in those territories could not defend their ownership in a court of law. Under such circumstances, it is understandable that most potential settlers chose not to embark on a physically hazardous venture that could well result in financial ruin. The king's officers, as it turned out, did not need to enforce the proclamation by raiding the offices of the settlement companies or harassing their investors. Washington need not have worried. The very fact that the king's edict was in force proved sufficient to effectively hamstring the colonists' efforts to settle the West.

And so, from year to year, frustration grew in British America. Investments were made; companies were formed; surveys—though technically illegal—were carried out. The money and resources were available, the potential settlers' hunger for land was palpable, and the trans-Appalachian West, now under British control, lay waiting. Everything was ready for the settlement of the western territories, and yet the settlement could not go forward. The line along the peaks of the Appalachian Mountains, drawn in a few brief sentences in the Proclamation of 1763, stood like an unbreachable dam against the rising pressure from the colonies. While that invisible line was in place, the American colonists, angry and frustrated though they were, could not move beyond the Atlantic Seaboard.

So things stood in the fall of 1774, as tensions between the American colonists and their king were reaching a boiling point. In December of the previous year, members of a secret society known as the Sons of Liberty had boarded the ships of the East India Company in Boston Harbor and dumped their entire cargo of tea into the bay. The Boston Tea Party, as it quickly became known, was a protest against the specific provisions of the unpopular Tea Act of 1773, but it was also a direct challenge to the right of British authorities to regulate commerce in the colonies. Parliament responded with a series of repressive measures aimed specifically at punishing the city of Boston and colony of Massachusetts but also at putting the other colonies on notice that acts of defiance would not be tolerated. Unfortunately for British authorities, these actions did not have the desired effect. Instead of putting an end to the disturbances, as the king and Parliament had hoped, the measures galvanized the colonists' opposition to the crown. To Americans they became simply "the Intolerable Acts."

In September 1774, each of the colonies (Georgia excepted) sent representatives to a Continental Congress in Philadelphia to decide on a common strategy in their fight against what they perceived as the king's tyrannical actions. In the end, the delegates met for only two months, and the decisions they reached were on the whole modest and conciliatory. And yet there is no denying the significance of the gathering, later dubbed the First Continental Congress: it was the first time the disparate American colonies had joined together to act as a single body. The debates that fall were long and inconclusive, but a new nation was slowly being born.

The Rights of Free Men

It was a time of great passions and great drama, and young Thomas Jefferson was swept up by the swirl of events. Having graduated from the College of William and Mary in 1762, he went on to study law and was formally admitted to the Virginia bar in 1767. Not content with pursuing a private career in momentous times, he entered into local politics and two years later was elected to the Virginia House of Burgesses, the colony's legislative assembly. In 1774, as the First Continental Congress was gathering in Philadelphia, Jefferson was a rising star in the legislature, as well as a fiery defender of the rights and freedoms of the American colonists. Though never a great orator, he was widely admired for his intellect, his encyclopedic learning, and his matchless literary gifts. Where others complained of unfair and discriminatory policies, he saw a clash of political philosophies, and where others spoke of economic interests and legal strategies, he spoke of the sweep of history and universal principles. He was the man who would turn the specific complaints of the disgruntled colonists into a grand vision of a new and just society that would arise in the New World.

Jefferson was not present at the First Continental Congress and did not participate in its long and sometimes fruitless debates. Instead, he composed a tract summarizing the grievances of colonists against the king and his government and had the Virginia delegates submit it to the Congress. Once adopted by vote, Jefferson hoped, it could be forwarded to London as a concise expression of the colonists' complaints and demands for redress. In the event, things did not go quite as planned: the colonies' representatives took their time with the tract and debated it at length but did not adopt it. Instead, they opted for a more limited and moderate remonstrance to the king.

Jefferson's first public intervention in the struggle with the king failed to produce the effect he had intended. Not only did the Congress refrain from adopting his proposal but many delegates were alarmed at his sweeping assertion of the inalienable rights of the colonists and the injustices of the king. And yet Jefferson's tract left an indelible mark on the political life of the colonies, for none could deny that he had raised the discussion to a whole new level. Instead of complaining about the revenues lost to Boston merchants and Virginia planters by royal action, Jefferson was speaking of the historical rights of Englishmen and the universal rights of all men and all societies. Two years later, when in a very different mood the delegates to the Second Continental Congress sought to justify their cause and present a new nation to the world, they looked no further: they chose Thomas Jefferson to compose the first draft of the Declaration of Independence.

A *Summary View of the Rights of British America*, as Jefferson's tract was called, touched on all aspects of the colonists' conflict with the king.[23] It condemned the Stamp Act and the Tea Act, the imposition of duties on the commerce of the colonies, and the declaration that resolutions of king and Parliament were supreme over acts passed by the colonies' legislative bodies. It defended the Boston Tea Party and fiercely denounced the retaliatory measures imposed by the crown, including the blockade of Boston Harbor. Those were the burning issues of the day, the ones that had brought the colonies together in united opposition to the king, and Jefferson accordingly gave them primacy of place in his protest against royal policies. Yet behind the immediate causes of the crisis and the growing rage of many colonists against the king lurked the unresolved problem of the western lands: the colonists were poised to settle them, but the king would not permit. It was this question that was most familiar to the son of Peter Jefferson and closest to his heart, and it was this question that elicited his boldest, most sweeping, and most eloquent rhetoric.

In demanding relief from the Proclamation of 1763, Jefferson made two distinct arguments, both of which challenged the king's right to dispose of his territories as he saw fit. The first was legal and historical, rooted in the ancient traditions of England itself—or at least Jefferson's interpretation of them. "Our Saxon ancestors," he wrote confidently, "held their land . . . in absolute dominion, disencumbered with any superior, answering nearly to the nature of those possessions which the feudalists term allodial." This was indeed the kind of argument one would expect from a young lawyer like Jefferson, seeking to make his mark in the world. Unlike feudal holdings, he explains, which were granted by one's superiors and ultimately by the king, an allodial title implied full unlimited ownership and the right to do as one pleased with the land.[24]

The feudal system, Jefferson explains, was unknown in England until it was introduced by William the Conqueror after his victory at Hastings in 1066. Since many of the great Saxon nobles were killed in the battle and subsequent uprisings, he divided up their estates among his followers as feudal holdings, granted by himself as King William I. In subsequent centuries, Jefferson continues, William's royal heirs did their best to expand the feudal system to cover more and more land: "A general principle, indeed, was introduced, that 'all lands in England were held either mediately or immediately from the crown.'" This, however, was mere trickery by "Norman lawyers." In truth, feudal holdings in England were few and far between, whereas for the most part English ownership followed "the Saxon laws of possession, under which all lands were held in absolute right."[25]

The same pattern, according to Jefferson, repeated itself during the settlement of British America. Since "America was not conquered by William the Norman, nor its lands surrendered to him," it followed that "possessions there are undoubtedly of the allodial nature." Yet once again it served the purposes of the king and his councillors to pretend otherwise and take advantage of the settlers, who, as Jefferson notes, "were farmers, not lawyers." Not knowing any better, the early settlers accepted land allocations from the king, in the mistaken belief that it was his right to grant them. This was not a problem as long as reasonable terms were set that did not interfere with the growth of the colonies. Now, however, with "the acquisition of lands being rendered difficult, [and] the population of our country is likely to be checked" because of the king's action, the fiction must be exposed: "It is time, therefore, for us to lay this matter before his majesty, and to declare that he has no right to grant lands of himself." For the land, according to Jefferson, belonged fully and unconditionally to the people who worked it.[26]

To the delegates of the First Continental Congress, and to the many colonists who purchased the printed treatise, Jefferson's argument was boldly radical, if not outright subversive. In the two centuries of English colonization of America, the right of the sovereign to grant his subjects lands in which to settle had never been questioned. Back in the sixteenth century, it was Queen Elizabeth who granted her servant Walter Raleigh all "the Newfound Land of Virginia," and although his efforts to settle the territory ended in spectacular failure, the precedent was set. From then on, the kings and queens of England, whether Stuart or Hanoverian, disposed of their American lands as they saw fit, and it never occurred to anyone to question the practice. To Europeans of the seventeenth and eighteenth centuries, who had lived in states governed by divine-right kings from time immemorial, it seemed like the most natural thing in the world. Few of them, if any, could even imagine an alternative arrangement.

Yet here was an obscure lawyer and member of the Virginia House of Burgesses who was not only questioning the practice but denying it outright. If Jefferson was to be believed, the very basis of land ownership in America, which had been in effect since the first colonists had settled in Jamestown, was null and void. It was nothing but a fraud concocted and perpetuated by the king and his lawyers to take advantage of the hardworking but naive settlers of the New World. Jefferson was not asking the king to correct misguided land policies. He was flatly denying the king's power to put in place any policies at all and calling him out as a power-hungry fraudster.

The delegates to the Continental Congress were no doubt impressed with Jefferson's reasoning, but they were also alarmed at its subversive implications. Only seven years before, Washington had asked a friend to keep his mild critique of the Proclamation of 1763 "a profound secret," and yet here was Jefferson openly denying not a specific rule but the king's very right to make it! In 1774, despite the rising tide of resentment in the colonies toward British rule, and despite an increasingly confrontational mood among the colonists, this was not a stance that most delegates were willing to adopt. And so they discussed Jefferson's proposal, then politely shelved it, and moved on to more practical items of business. It would be another two years, years in which the bonds connecting the colonists to their king would increasingly fray and then rupture, before the delegates to the Second Continental Congress were willing to adopt such dangerous resolutions.

But if Jefferson's argument that the colonists had an allodial right to their land was too radical for the Continental Congress, to Jefferson himself it seemed not radical enough. True, the argument was bold, and its broad historical sweep uprooted hundreds of years of royal prerogatives in both England and America. And true, its final conclusions fully voided the king's rights to grant or withhold American lands as he had in the infamous Proclamation of 1763. Nevertheless, for all of Jefferson's denigration of the monarch's clever lawyers, the argument of the *Summary View* so far was itself a legal argument. And a legal argument, the young lawyer knew well, inevitably begets another legal argument, often equally compelling but leading to opposite conclusions.

Jefferson's claims stood or fell on his interpretation of Saxon law and traditions and their place in the legal order established by their Norman conquerors. One can easily see how a jurist in the king's service could challenge Jefferson's sweeping claims and bring a treasure house of documents from the Royal Archives to support his position. Yet even if we grant that Jefferson's version of history was correct, other questions arise. Why, for example, should Saxon law prevail over all others? Why should the legal reforms introduced by the Normans not be given at least equal weight to the Saxon traditions? Or if it is the antiquity of the code that counts, why not go back to the Britons, who lived under Roman law before they were vanquished by the Saxons? Jefferson may well have had good answers to all these questions, but the fact remains: arguments based on history and legal interpretation most often invite counterarguments based on history and legal interpretation. They are never conclusive.

And there is also this: if one lives by the precedent of history, one might also die by the same precedent. It might be true, as Jefferson insisted, that Saxon traditions guaranteed the settlers full rights of ownership over their land. But it was certainly true that English traditions guaranteed that subjects must obey their king. And what could the ardent revolutionary Jefferson say to that? The truth is that for a radical thinker such as Jefferson, historical precedent was a problematic basis for an argument and was quite likely a trap. Arguing for radical change based on established precedents can easily end up undermining one's own position. And so, to clinch his case for the colonists' rights, Jefferson needed something more: an argument based not on the contingencies of history but on the very laws of nature.

And that indeed was the heart of Jefferson's second argument against the king's hold on American lands. "Our ancestors," he wrote, "before their emigration to America . . . possessed a right which nature has given to all men . . . of going in quest of new habitations and there establishing new societies, under such laws and regulations as to them will seem most likely to promote public happiness."[27] The right of colonists to settle and establish new societies, then, is not dependent on particular traditions, such as the Saxon laws of land-ownership. They are, rather, the inherent rights "that nature has given to all men," regardless of their origin or history.

It follows, Jefferson explains later in the *Summary View*, that "all the lands within the limits which any particular society has circumscribed around itself are assumed by that society and subject to their allotment only."[28] The implication for the American colonists was clear: the minute they settled down and marked out a certain territory for their use, all the land in that territory became theirs by natural universal law, and they were free to divide it up as they saw fit. This, Jefferson explains, could be done by the settlers collectively or by an elected legislature to which they delegated this power. But if no such systematic division of land were undertaken, then Jefferson is very clear: "Each individual of the society may appropriate to himself such lands as he finds vacant, and occupancy will give him title."[29]

This, needless to say, was music to the ears of the colonists who were eager to set off to the West and settle the lands beyond the Appalachians. Here Jefferson was not only denying the king's right to interfere with the settlement of America; he was denying *anyone's* right to interfere and boldly asserting that any land enclosed by the settlers was theirs by natural universal right. If Jefferson had his way, the floodgates would finally open, and the settlers would pour over the mountains and freely settle the plains beyond, confident that the lands were theirs by right.

There was, to be sure, the minor caveat that the lands must be "vacant," which, given the stubborn presence of the native tribes, they clearly were not. According to the *Summary View*, the settlers were taking possession of "the wilds of America," where no man had previously possessed any land, and the land was consequently vacant and ready for its new owners. While this was undoubtedly a supreme exercise in wishful thinking on Jefferson's part, it was also well in line with the attitudes of early settlers, as well as his own land speculator father. And if Jefferson was seeking a respectable philosophical grounding for his views, he likely found it in the writings of his hero, John Locke, who argued that landed property derived from its agricultural cultivation. The Indians, Locke wrongly claimed, lived in the "wild woods and uncultivated waste of *America*" and therefore had no claim to its land. This was a reckless mischaracterization of native practices, and it led to disastrous consequences for American Indians. But for Jefferson and other advocates of western expansion, it was a convenient fiction, as it transformed the thoroughly populated West into a vast and empty expanse awaiting its settlers.[30]

It is not that Jefferson literally believed that the lands of the West were devoid of inhabitants. Indeed, there can be no doubt that Jefferson was well aware of the longtime presence of native people in North America. His own *Notes on the State of Virginia* include a sympathetic and even admiring account of the Indians, as well as a list of the different tribes in the area of Virginia and their territories. And years later, as Meriwether Lewis and William Clark were preparing for their western expedition, he instructed them to gather as much information as possible on the native peoples they encountered on their journey. But when it came to the settling of the West, the Indians, though undeniably present, simply did not count. They could, if they chose, join the mass of white settlers, and Jefferson hoped they would; or they could refuse, in which case they would be removed or destroyed. Either way, they had no claim to the lands they had inhabited for centuries. The lands of the West were, accordingly, entirely empty and awaiting their settlers.[31]

Jefferson's first argument, anchored in the traditions of Saxon law, showcases his broad historical understanding and his sharp legal mind in arguing against the king's right to American lands. It is the argument of a brilliant and unusually broad-minded lawyer, which is indeed what Jefferson was at the time and had been since his admission to the bar seven years previously. In his second argument, however, Jefferson transcends his legal roots and points the way toward his, and America's, future. In his general argument that people have the right to determine their own fate; in his unusual emphasis on the importance of promoting public "happiness"; and most of all in his insistence that all humans possess universal God-given rights that take precedence over manmade laws and regulations, we can already perceive the future author of the Declaration of Independence. It would be another two years before Jefferson wrote that "all men are created equal, that they are endowed by their Creator with certain unalienable Rights, that among these are Life, Liberty and the pursuit of Happiness." But the seeds of that famous declaration are already present in his bold defense of the rights of the colonists in A Summary View of the Rights of British America.

An Empire of Liberty

In 1774, Jefferson was at the very beginning of a career in public life that would last for more than half a century. In the following years, he would become chief author of the Declaration of Independence, and one of its signers; the governor of Virginia; a leading delegate to the Continental Congress; minister to France, America's foremost ally; secretary of state in George Washington's first administration; vice president under John Adams; and president of the United States from 1801 to 1809. After his retirement, he became the moving spirit behind the founding of the University of Virginia as well as its chief architect. He remained intimately involved in national political life until his death in 1826 at the age of eighty-three.

During those years Jefferson was constantly occupied with an endless stream of issues that arose during the crisis of the revolution and the early decades of the American republic. Military strategy, international treaties, the states' public debt, the location of the federal capital, piracy on the high seas, state rights versus federal authority, trade policy, neutrality rights, and treason by his own vice president are but a small sample of the incessant flood of challenges and crises that required his attention and energy at different times. And yet through it all, despite all the twists and turns of a decades-long political life in a rapidly changing America, the main themes he expressed in the *Summary View* back in 1774 run like a thread.

Even as governor of Virginia in 1779–80, during the bitter years of the Revolutionary War, Jefferson would not divert his eyes from the West. He enthusiastically supported the western campaigns of the frontier leader George Rogers Clark, who sought to lure the French settlers along the Mississippi into alliance with the United States instead of with their nominal overlord, Great Britain. The Frenchmen, who had never reconciled themselves to the loss of New France in the French and Indian War, were happy to oblige, especially after Clark showed his mettle by capturing the British outposts of Kaskaskia and Vincennes. In congratulating Clark, Jefferson wrote stirringly that his success "will add to this Empire of Liberty an extensive and fertile Country, thereby converting dangerous Enemies into valuable friends."[32] For it was in the vast expanses of the West, Jefferson knew, as far removed as possible from the corrupting influence of "civilized" Europe, that freedom would shine and the republic take root.

Jefferson's greatest and boldest stroke for western expansion came in 1803, during the first term of his presidency. Two years before, Napoleon Bonaparte, First Consul of what was still the French Republic, had pressured his Spanish allies into ceding their North American territories west of the Mississippi to France. Napoleon had grand designs for re-forming the French North American Empire that had been lost four decades previously, but it was not long before he realized that such dreams were impracticable. The grand army he had sent to reassert French control of Saint-Domingue was decimated by disease and soundly defeated by former African slaves, who called the territory Haiti. Meanwhile, a new war was looming against Great Britain, and with the mighty Royal Navy threatening all communications with the New World, the future emperor concluded that the colony of Louisiana would not be a feather in his cap but a drain on his resources.

As it happened, Napoleon's desire to unload his budding American empire coincided with Jefferson's interest in acquiring New Orleans as an essential outlet for American goods shipped down the Mississippi. Having no inkling that Napoleon was willing to dispose of his American domains, he sent his fellow Virginian James Monroe to negotiate a fair price for the port city. But when Monroe arrived in Paris, Napoleon's foreign minister, the Prince de Talleyrand, had a different proposal for him: for $15 million, he would sell the Americans not only New Orleans but the entire vast territory of Louisiana.[33] The sum Talleyrand demanded was $5 million more than Monroe was authorized to spend, and as for the territory, no one knew precisely where its boundaries lay or how large it actually was. Yet Monroe, knowing Jefferson's priorities, did not hesitate: he guaranteed the transfer of the funds, signed the contract, and closed the deal. "You've made a good deal for yourselves," Talleyrand reassured the nervous American negotiators, and so they had. With an area of 883,000 square miles, Louisiana nearly doubled the territory of the United States. And it sold at less than three cents per acre.[34]

The Louisiana Purchase was Jefferson's master stroke for the opening up of the western territories, and he immediately followed it with another. In 1804, he sent out an expedition to explore the newly acquired territory, map and survey it, and report on its plants and animals as well as on its Indian inhabitants. To lead the expedition he chose his personal aide, Captain Meriwether Lewis (1774–1809), and his friend Lieutenant William Clark (1770–1838)—younger brother of George Rogers Clark, western hero of the Revolutionary War. Setting out from St. Louis in May 1804, the expedition made its way up the Missouri River, across the Rocky Mountains, and down the Columbia River to the Pacific coast. It then retraced its steps and arrived back at its starting point in September 1806, turning the expedition's leaders into overnight national heroes.

The Lewis and Clark expedition did, to be sure, gather vital information that was later used to settle the territory with white colonists. But its impact on the young United States went far beyond such practicalities. The expedition inspired the American imagination and filled it with visions of the immeasurably vast, unspoiled, and untrammeled open spaces of the West, spaces that were waiting to be claimed and settled. The myth of the American West, with its intrepid settlers, wild Plains Indians, immense buffalo herds, clear streams, ferocious grizzlies, and wide-open "big sky," had its origins in the Lewis and Clark expedition.

As president, Jefferson doubled the territory of the Union and opened up much of the continent for settlement, yet his hunger for western lands went even beyond that. In the very first year of his presidency, while mulling over the acquisition of New Orleans, he was already envisioning the future expansion of the United States. Writing to James Monroe, he acknowledged that current conditions and interests "may restrain us within our limits." Nevertheless, he continued, "it is impossible not to look forward to distant times, when our rapid multiplication may expand beyond that limit and cover the whole Northern, if not the Southern continent."[35] Even while angling to get hold of a single city, Jefferson was envisioning the immense ultimate spread of the republic. Writing decades before the term *manifest destiny* was coined, Jefferson was already confident that the United States was destined to occupy an entire continent, if not two.

It is not that Jefferson was unaware that the western territories— not to mention the two continents of North and South America— were already peopled with native tribes. Ever curious, he had studied the native languages, and in his *Notes on the State of Virginia*, he wrote thoughtfully and even admiringly about the Indians, their customs, and his own personal experience with them. Before sending Lewis and Clark on their long journey, he instructed them to gather as much information as possible on the tribes they encountered along the way.[36] Yet when it came to the peopling of the West, the native people simply did not count for Jefferson. As far as he was concerned, the western lands were entirely empty and ready for settlement by white colonists. And if the facts on the ground did not fit with this idealized vision, then it would be the facts—not the vision—that would have to give way.[37]

An Inconvenient Presence

Reconciling the undeniable presence of native people throughout North America with the vision of an empty land awaiting its European settlers required considerable mental gymnastics, but Jefferson was up to the task. For even as he praised the courage and manliness of the Indians in his writings and argued that they were fully the equal of Europeans in vigor and intelligence, he also made it clear that they were a doomed race whose time on Earth was drawing to a close. In what is perhaps the most famous anecdote in the *Notes on the State of Virginia*, Jefferson tells the story of the Mingo chief Logan, who took bloody revenge after his family was massacred by white settlers. Jefferson might have taken the opportunity to reflect on the genocidal brutality of the settlement of the West and perhaps reconsider the justice of his expansionist vision. But that is not at all the lesson he took from the story. In his recounting, Logan is a tragic hero, bravely but hopelessly facing preordained and inescapable doom. "There runs not a drop of my blood in the vein of any living creature," Logan laments, foreshadowing what Jefferson believed would soon be the fate of all American Indians. "Who is there to mourn for Logan?—Not one."[38]

Jefferson, to be sure, did not deny that Logan had been the victim of an injustice, which was perpetrated by a particularly murderous white man named Cresap. But he is also careful to absolve the majority of settlers of any culpability in the crime. The demise of the Indians, in Jefferson's telling, is not the result of intentional acts by settlers or policy decisions by the US government. It is simply preordained, a manifestation of the universal law that the wild man will give way to civilized man.[39] "Their history will terminate," Jefferson wrote confidently in an 1803 letter, assigning no responsibility for this inescapable eventuality. As far as he was concerned, there was nothing anyone could do about it.[40]

This was unfortunate for the Indians, and Jefferson was happy to concede as much. But it also solved what to him was a key problem: since the Indians were a dying race whose demise was assured, Jefferson considered himself perfectly justified in viewing the vast lands they inhabited as effectively empty and open to white settlement. This, to Jefferson, was the general principled response to the inconvenient presence of Native Americans in America. And Jefferson, after all, was a man who liked to think in general principled terms. Yet while this preserved and justified Jefferson's vision for America as an empty land, it was also at best a long-term solution that did nothing to address the crisis at hand. In the short term, the Indians were not at all gone but very much present in their lands. Many of them, furthermore, seemed fully ready to defend those lands, quite unaware of the imminent termination of their history. Meanwhile, waves of impatient European settlers were flooding into the West, clearly disinclined to wait until the Indians disappeared of their own accord. To the US government, these conflicting trends posed a thorny question: What must be done to clear the West of the Indians and open it to white settlement?

Jefferson had an answer. The best way to "disappear" the Indians and open up their territories was to absorb them into the mass of white settlers staking their claim to the land. This, he argued, would be in the Indians' best interest because if they continued in their traditional way of life, they would soon become extinct. Their best course, in fact their only hope of physical survival, was to surrender their unique identity and assimilate into American society. "The ultimate point of rest & happiness for them," Jefferson wrote benevolently in 1803, "is to let our settlements and theirs blend together, to intermix and become one people, incorporating themselves with us as citizens of the US." Of all "the various ways in which their history may terminate," he assured his correspondent, Benjamin Hawkins, "this is the one most for their happiness."[41]

To hasten the process of transition from nomadic to settled life, Jefferson made it US policy to encourage the spread of the "civilized" practices among the Indians. "I am happy to inform you," he wrote hopefully to Congress in 1801, "that the continued efforts to introduce among them the implements and the practice of husbandry and the household arts have not been without success."[42] Once the Indians had adopted farming, he reasoned, they would require much less land to survive and so would be willing to surrender it. European settlers, meanwhile, would require ever more land to distribute to their ever-increasing numbers. This, in Jefferson's opinion, was a "coincidence of interests," in that the transfer of territory from the oversupplied Indians to the undersupplied settlers would be "for the good of both."[43]

Yet optimistic as all this sounds, Jefferson was nothing if not a realist. Waiting for the Indians to adopt European ways would require putting his plans for the settlement of the West on hold for an indeterminate length of time, which neither he nor the would-be settlers were willing to do. Furthermore, he may have suspected that the benevolence of his plan for cultural erasure would be lost on the Indians, who might irrationally choose to cling to their old way of life. And so, to ensure the swift transfer of native land to the US government, he included some less idealistic incentives as well, to be used as necessary.

The first of these was devious. In an 1803 letter to future president William Henry Harrison, who was serving as governor of the Indiana Territory, Jefferson proposed to "push our trading houses" on the Indians. The purpose, he explained with impressive candor, was "to see the good & influential individuals among them run in debt, because we observe that when these debts get beyond what an individual can pay, they become willing to lop th[em off] by cession of lands."[44] And if economic extortion failed to separate the Indians from their land, there was also the second incentive: the ever-present threat of violence.

As a bloody history of war over several centuries attests, the threat of using violence to uproot and kill lurked just beneath the surface of Europeans' dealings with Native Americans. And Jefferson, in this regard, was no exception. Already during the Revolutionary War, as governor of Virginia, he had suggested extreme measures for dealing with recalcitrant Indians. If you choose to move against "these Indians," he wrote western commander George Rogers Clark, "the end proposed should be their extermination, or removal beyond the lakes or Illinois river."[45]

Decades later, the drumbeat continued. In 1803, now president Jefferson wrote Governor Harrison, "Should any tribe be fool-hardy enough to take up the hatchet . . . the seizing of the whole country of that tribe & driving them across the Mississippi . . . would be an example to others."[46] In 1807, faced with the pan-tribal movement led by Shawnee war leader Tecumseh and his brother, the prophet Tenskwatawa, Jefferson warned "that if ever we were constrained to lift the hatchet against any tribe, we will never lay it down until that tribe is exterminated, or driven beyond the Mississippi,"[47] and in the midst of professing his friendship to the Indian nations in 1809, he also reminded them that "the tribe which shall begin an unprovoked war against us, we will extirpate from the Earth, or drive to such distance that they shall never again be able to strike us."[48] "In war," he conceded to Secretary of War Henry Dearborn, "they will kill some of us," but "we shall destroy all of them."[49]

All in all, over his long public career Jefferson proposed a range of solutions to his intractable "Indian problem." Ideally the Indians would simply melt into the sea of Europeans that was flooding North America and disappear within them. That was undoubtedly his preferred solution and the one most consistent with his views of justice and of himself as an enlightened man. He also believed that, in the long run, it would inevitably come to pass. But if the tribes proved recalcitrant or the process simply too slow, he did propose other ways of dealing with the Indians: economic blackmail, removal beyond the Mississippi—and, if all else failed, extermination.[50] All these different options shared a core fundamental assumption: the Indians, as Indians, had no claim on the land and no place in the new America. The territory, Jefferson insisted, was vacant. And if it would appear to an outside observer that there were already people living in it, Jefferson stood ready to correct them. By the time they looked again, he would make sure, the native people will have disappeared. The continent of North America was just as he imagined it—a vast untrammeled empty space, ready to be settled and farmed.

"Those Who Labour in the Earth"

Toward the people who stood ready to replace the Indians in the American West, Jefferson's attitude was very different. Indeed, his admiration for the yeoman farmers who would, he imagined, settle the West knew no bounds. They were new men like none other, who would build a new land like none before. They were enterprising, virtuous, law-abiding—and most importantly, freedom loving, the backbone of the American republic. This lofty vision of the new Americans stirred Jefferson to flights of rhetoric: "Those who labour in the Earth," he wrote in the *Notes on the State of Virginia*, "are the chosen people of God, if ever he had a chosen people, whose breasts he had made his peculiar deposit for substantial and genuine virtue. It is the focus in which he keeps alive that sacred fire, which otherwise might escape from the face of the Earth." In contrast, the moral condition of city dwellers was as depraved as that of their rural brethren was exalted: "The mobs of the great cities," Jefferson wrote with undisguised contempt, "add just so much to the support of pure government, as sores do to the strength of the human body."[51]

And if this, as Jefferson believed, was the case in any human society that had ever existed, it was immeasurably more so in America. Because it was here, in a free republic planted in the vast expanse of an entire continent, that rural virtue was given free rein. "We have now lands enough to employ an infinite number of people in their cultivation," he wrote John Jay in 1785, as American settlers were beginning to pour into the lands beyond the Appalachians. "Cultivators of the earth," he continued, "are the most valuable citizens. They are the most vigorous, the most independent, the most virtuous, & they are tied to their country & wedded to it's liberty and interests by the most lasting bonds."[52] They are, in other words, the natural bearers of the republican ideal of the new United States.

Jefferson's faith in the moral purity of rural life and the republican virtue that stems from it never wavered. Nearly three decades after he waxed lyrically to John Jay on the noble qualities of the "cultivators of the Earth," he was still explaining to John Adams why American farmers were natural republican citizens. The oppressive regimes of the Old World, he wrote, were all well and good for "the Man of the Old World, crouded within limits either small or overcharged, and steeped in the vices which that situation generates." But in America things were very different because "here every one may have land to labour for himself if he chuses" and is consequently "by his property . . . interested in the support of law and order." Such men have no use for European style tyrannies because they "may safely and advantageously reserve for themselves a wholesome control over their public affairs, and a degree of freedom" that in the hands of European city dwellers "would be instantly perverted to the demolition and destruction of everything public and private."[53] As a cultivator of the earth, by his own choice and with his own industry, the new American man was a natural republican citizen who would not tolerate the heavy hand of a tyrannical government.

And this is precisely why Jefferson forever insisted that America was a vast and empty virgin territory and would never acknowledge any other truth. It was this vastness that would allow it to accommodate "an infinite number of people." It was its emptiness that would allow this multitude to settle, farm, and cultivate the land. And it is this new and numerous people, spread throughout the continent and each cultivating their own plot of land, that would be the backbone of the republic. Whether the United States, a young republic in a world dominated by monarchies, would survive and thrive was an open question in Jefferson's day. But if it were to prosper, he believed, it would be thanks to those freedom-loving, enterprising, hardworking yeomen farmers of the West. Through their work on virgin soil, they would write their story—the story of a great, new, free republic.

The American West was, for Jefferson, a vast and all but limitless space, stretching, as it did, across an entire continent. It was also, for all intents and purposes, an empty space, at least as far as humans were concerned. When some humans were inconveniently found to inhabit this "empty" land, he used all necessary means to make sure that it was indeed empty. And it was a uniform space, governed entirely by American laws and customs and soon to be inhabited by American settlers. The entire continent, Jefferson wrote hopefully to Monroe in 1801, would be covered by "a people speaking the same language, governed in similar forms & by similar laws." Diversity, which he called "a blot or mixture on that surface," would not be allowed.[54]

Jefferson had reduced the space of an entire continent, including its natural topography and human inhabitants, into a pure abstraction. America was for him an empty, uniform, limitless, space—a blank slate on which enterprising American settlers could employ their industry, write their own story, and create a new nation like no other. Two years before his death, Jefferson reflected that whereas Europeans were forever constrained by the weight of historical and legal precedents, America "presented us an album on which we were free to write what we pleased."[55] He was referring specifically to the American Revolution, which succeeded far more than the English Revolution of the seventeenth century because, as he writes, it "commenced on more favorable ground."[56] But what was true of America in general was even truer of the vast expanses of the American West, which for Jefferson was the physical expression of this very ideal: here was America as a blank slate, a tabula rasa or "album," on which its new people, unconstrained, would write the story of a new nation.

Limitless, empty, and uniform, the American West as Jefferson saw it was the very embodiment of Newtonian absolute space. Newton, it will be recalled, insisted that empty space existed in and of itself, before it was filled with any objects or became the setting for any actions. It was boundless in every direction, had no center or periphery, and, being completely empty, was also uniform—everywhere the same. It was precisely these characteristics of space—its boundlessness, emptiness, and uniformity—that, according to Newton, made it possible for God to create the world within it as he saw fit. Descartes's world was entirely full of matter and operated like a perfect machine in which object constantly pushed against object, leaving no room for outside intervention. But Newton's empty world, existing as it did in absolute uniform space, allowed for freedom—freedom to create the sun, the moon, the planets, and the stars and to imbue the universe with the mysterious power of gravity.

Needless to say, the freedom to create a universe is God's and God's alone. And yet, as Bishop Stillingfleet pointed out to Locke, the freedom of action granted to God in Newtonian space applies to humans as well.[57] Within this great and uniform emptiness, they too may attempt their own humble creations, though on a human rather than divine scale. Unconstrained by the all-encompassing Cartesian machine, humans are free to exercise their talents, their industry, and their beliefs and shape the world in their own small way.

"We rush like a comet into infinite space," warned the exasperated Federalist leader Fisher Ames in 1803, conjuring the unbounded and empty Newtonian universe to denounce Jefferson's westward drive.[58] And Jefferson would not have disagreed, for that was precisely how he envisioned the West. The men of Europe, and to some extent the men of the eastern colonies, could never be fully free: everywhere, like the inhabitants of a "full" Cartesian world, they are constrained by history, traditions, and the push and pull of their fellow men. The American West, in contrast, in its vastness, its uniformity, and most of all its glorious emptiness, is the place of unconstrained freedom. In this vast expanse, men would settle, cultivate the land, and create a free society such as the world had never seen.[59]

This grand vision for a new land and a new nation remained with Jefferson from the earliest days of his public life to the very end. Already in the *Summary View* of 1774, he had insisted on the rights of settlers to form their own society and their own laws in the "vacant" western territories. And in his very last years, he once again referred to the New World as an "album" in which Americans were "free to write what we pleased." It was an inspiring promise of freedom in America, and Jefferson never wavered from it. And it was based on

his conviction that the American West, at its core, was absolute Newtonian space.[60]

The American Newton

As citizens of the twenty-first century, we find it hard to believe that
philosophical and scientific debates on the nature of space would have
any impact whatsoever on social and political life. Recent ground-
breaking scientific discoveries such as the detection of gravity waves
and the Higgs boson particle, as well as debates over the merits of
string theory or the true meaning of quantum mechanics, have profound implications for the nature of our universe and our place within
it. But it is all but impossible to imagine a political leader, on the
right or on the left, taking a position on any of those questions, not
to mention founding a political program based on one side of these
issues or the other. Some scientific issues, to be sure, have not lost
their political charge even today. Darwinian evolution, the science of
climate change, and most recently immunology still take center stage
in our present-day culture wars. Yet when it comes to the physical
and mathematical sciences, they seem to have become fully insulated
from the rough and tumble world of politics and ideologies.

Jefferson, however, was a child not of the twenty-first but of the
eighteenth century, commonly known as the age of Enlightenment.
And at that time things were different. The leading Enlightenment
philosophers and men of letters, men such as John Locke and David
Hume in Britain or Voltaire, Denis Diderot, and Jean-Jacques Rousseau
in France, held sharply conflicting views on many issues and argued
incessantly with one another. But they all agreed that they were living
in a new age of reason, the first since the dawn of humanity. Since the
beginning of time, they believed, humans had wallowed in ignorance
and superstition, a condition that led inevitably to poverty, misery,
and oppression. Now, at long last, a new age had dawned, an age in
which reason would vanquish ignorance and men would be lifted
from their misery and led to a just and prosperous future. Leading
the way forward was the new science, and in particular Newtonian

science, which had deciphered the mysteries of the universe and reduced them to simple, rational, mathematical equations. If the same methods could be applied to other fields—scientific, social, or political—then one could expect similar breakthroughs there as well. A brilliant new day would assuredly dawn.

For the men of the Enlightenment, then, Newtonian science was not just a technical mathematical account of the physical world, on a par with modern scientific theories like relativity and quantum mechanics. It was, rather, a shining beacon showing the way to an enlightened, peaceful, and prosperous future anchored in the rule of reason. Consequently, adherence to Newton's methods and principles was more than a scientific matter: it was, rather, a statement of allegiance to the values and the promise of the Enlightenment. And so, if that secular age of reason nevertheless required a patron saint, he was, indisputably, Isaac Newton.

It is probably safe to say that in all of colonial America and the early United States there was no greater admirer of Newton than Thomas Jefferson. While it is likely that the name of the celebrated Englishman had penetrated even to rural Virginia, where Jefferson spent his youth, the future president's first study of Newtonian science took place during his years as a student in the College of William and Mary, between 1760 and 1762. There, under the guidance of his teacher William Small (1734–75), Jefferson was introduced to the world of Enlightenment science and philosophy. He studied Newton's "method of fluxions," more commonly known today as the calculus, and worked his way through the *Principia*—surely making him the only American president to have read the work. Later, during his years as secretary of state, Jefferson hung a portrait of Newton in his office, before transferring it to Monticello, where it still hangs alongside the portraits of Sir Francis Bacon and John Locke in what became known as "the gallery of the immortals." As a final act of devotion, Jefferson managed to acquire Newton's death mask, one of several taken upon his death in 1727.[61]

Jefferson's admiration for Newton was not just a matter of his personal high regard for the Englishman's scientific accomplishments. It was, rather, a public stance, openly presented to anyone visiting his office in New York or his home in Monticello. By displaying Newton's portrait, not to mention his death mask, Jefferson was proclaiming his allegiance to the ideals of the Enlightenment and his belief that through reason men will improve themselves and their condition in the world. And while this promise of the Enlightenment had been repeatedly frustrated in the Old World, in America, he believed, it will finally be redeemed. Unencumbered by the prejudices of the Old World and free to exercise reason to the fullest degree, the men of America will create the kind of society that Enlightenment thinkers had promised: just, prosperous, and, above all, free. For Jefferson, in other words, the portrait of Newton stood for the promise of America.

This is why, as has often been noted, Jefferson's stirring words in the Declaration of Independence are imbued with a Newtonian outlook. His invocation of "the Laws of Nature and of Nature's God" immediately bring to mind a Newtonian world governed by universal uniform laws that have their origins in God. So does his bold declaration that "all men are created equal" and "are endowed by their Creator with certain unalienable Rights." The world he brings forth here is a world of universal equality and uniformity, which—just like Newton's universe—is permeated by all-encompassing principles derived from God. The United States, Jefferson promises, will be a Newtonian land in which universal principles will, for the first time, find their full expression and create a perfect society.[62]

And just as the Declaration of Independence was the document that laid out a utopian society based on universal Newtonian principles, the American West was the space in which this society will take root. The West, to Jefferson, was the embodiment of Newtonian space—seemingly limitless, entirely (so he insisted) empty, and completely uniform, free of history, hierarchy, and traditions. It was as close as one could get to the empty and uniform space of Newton's *Principia*, which Locke had identified as the space of freedom. It was, accordingly, precisely the space in which a new society, based on universal principles and the free exercise of reason, could flourish.

But if the American West was indeed an absolute Newtonian

space, how could one take possession of it? In a completely empty space that has no center and (for all practical purposes) no beginning and no end, where every point is in principle the same as any other, how does one divide it up and differentiate one location from another? Newton himself had already confronted a similar problem when dealing with the absolute space of his world system and had come up with the solution. Absolute space, for him, was the mathematical space in which the curves of analytic geometry take shape. This space, as he shows in the *Enumeratio*, is defined by the rectilinear coordinates x and y in a plane, or x, y, and z in the case of a three-dimensional space. From these axes emanates a uniform grid of straight lines that intersect one another at right angles. Thanks to this all-pervading grid, every point in mathematical space is uniquely defined, distinguishing it from all others. The grid, in other words, takes the amorphous, centerless, and diffuse mathematical space and turns it into an orderly regimented space, in which each point is unique and graphs can be traced.[63]

Newton's solution was, as we have already seen, a powerful mathematical tool that played a crucial role in the advancement of analytic geometry. It was also essential for the related field of analytical mechanics, which flourished in the eighteenth century and which we would probably call mathematical physics. And yet Newton's gridding of space was the solution of a sophisticated mathematician, one who moved easily between the physical world of material objects and the abstract world of mathematical ones. It was an approach that allowed him to graph complex algebraic formulas and classify them in ways that were meaningful to mathematicians. But what, one may ask, does this have to do with Jefferson's mission of settling the American West? What did Jefferson know or care about Newton's higher mathematics?

Thomas Jefferson, Mathematician

As it happens, a great deal. For long before Jefferson became a Founding Father of the United States, before he was elected president, acquired

Louisiana, and sent Lewis and Clark on their journey, before even the hectic days of 1776 when he authored the Declaration of Independence, young Jefferson's ambition was to be something else entirely: a mathematician. "When I was young mathematics was the passion of my life," Jefferson wrote to his friend, the lawyer and politician William Duane, when he was approaching his seventieth year and musing on a road not taken.[64] "It was ever my favorite" study, he recalled in an 1811 letter to Benjamin Rush: "We have no theories there, no uncertainties remain in the mind; all is demonstration and satisfaction."[65] And so, despite his long career in public life, Jefferson never abandoned his youthful fascination with mathematics but returned to it again and again.

The man who introduced him to the mysteries of higher mathematics was his college mentor, the Englishman William Small, who was an Enlightenment scholar and polymath in his own right. Jefferson and Small quickly forged a close bond, which Jefferson would warmly recall decades later. Small, he wrote to H. L. Girardin in 1815, "was to me as a father. To his enlightened and affectionate guidance I am indebted for everything."[66] Under Small's guidance, Jefferson studied Euclidean geometry before moving on to Newtonian mathematics. He learned Newton's calculus from William Emerson's textbook, *The Doctrine of Fluxions*, and then put it to use by studying Newton's own *Principia* and *Opticks*.[67] Nor did his interest in mathematics abate after he left college and embarked on a legal and then political career. In 1789, while serving as minister to France and closely observing the dramatic events in Paris that would lead to the outbreak of revolution, he yet found time to report to Joseph Willard, president of Harvard College, on the latest mathematical development. "A very remarkable work is the 'Mechanique Analytique' of Le Grange, in quarto," he wrote of Joseph-Louis Lagrange's seminal work that would transform both mathematics and physics. "He is allowed to be the greatest mathematician now living, and his personal worth is equal to his science," Jefferson continued, before adding an accurate and well-informed summary of the work: "The object of the work is to reduce all the principles of mechanics to the single one of equilibrium, and to give a single formula applicable to them all. The subject is treated in the algebraic method, without diagrams to assist in conception."[68] Lagrange himself could not have said it better.

Thirty-five years later, Jefferson was still keeping abreast of the latest developments in mathematics. "The English generally have been very stationary in later times, and the French, on the contrary, so active and successful," he wrote to Professor Patrick K. Rogers of the College of William and Mary, noting the relative decline of the English mathematical school and the ascendancy of the French. He then went on to reference the work of the most prominent French mathematicians of the day: "Lacroix in mathematics, Legendre in geometry . . . to say nothing of Monge and others, and the transcendent labors of Laplace." Not only was Jefferson knowledgeable about latest and most influential work but he was also closely following the latest professional trends. Cambridge mathematicians, he explained, feeling that they were falling behind their continental colleagues "due to their long-continued preference of the geometrical over the analytical methods," had decided to reverse course and adopt the French approach. In particular, "they have given up the fluxionary [i.e., Newtonian], for the differential [i.e., Leibnizian]calculus."[69]

Jefferson's account of the state of mathematics in 1824 is entirely accurate. French mathematics, with its emphasis on analytical over geometrical methods and routine use of the Leibnizian calculus, was ascendant, led by the very mathematicians Jefferson cites. In Cambridge, meanwhile, a group calling itself the "Analytical Society" had formed around Charles Babbage and John Herschel and was working to revive the fortunes of English mathematics by adopting French methods. What is remarkable is that two years before his death Jefferson was still keeping abreast of the field that had captured his heart in his youth and could confidently report on its leading practitioners and their work and even on the shifting balance of power between competing national schools. It is safe to say that there are very few other eighty-one-year-olds capable of such a feat, and no ex-presidents.

Jefferson's correspondence from the intervening years is sprinkled with allusions to mathematics, which was forever tempting him away from his public duties. In December 1788, even as great events were underfoot just outside his window, Jefferson wrote from Paris to the celebrated radical Tom Paine not on the gathering storm of revolution but on the ideal shape of the arch of a bridge. Citing the work of Italian mathematician Lorenzo Mascheroni, Jefferson recommended the catenary—the shape of a hanging chain—a complex curve whose equation had quite recently been determined by Johann Bernoulli with the aid of the differential calculus.[70] Around the same time, and with a similar eye to practical mathematics, Jefferson became obsessed with determining the ideal shape of a moldboard—the surface of a plow that overturns the earth. When after a decade of investigations he published his answer in the *Transactions of the American Philosophical Society*, he argued that the correct answer could be derived from Newton's discussion of "solids of least resistance" in the *Principia*.[71]

Even during the late 1790s, when Vice President Jefferson was leading the opposition to President John Adams's Federalists, he still took the time to plot a detailed course in mathematics for fellow Virginia planter William Green Munford. He recommends Euclid, and "some of Archimedes," but most importantly trigonometry, which is most valuable to every man. Speaking as the true son of a surveyor, Jefferson assures Green that "there is scarcely a day in which he will not resort to it for some of the purposes of common life." Algebra, logarithms, and the extraction of square and cube roots are also useful, in Jefferson's opinion. "All beyond these," including conic sections, spherical trigonometry, algebra beyond quadratic equations, and "fluxions" (i.e., the calculus) "is but a luxury." Such studies, he warned his friend, are "a delicious luxury indeed; but not to be indulged in by one who is to have a profession to follow for his subsistence."[72] Jefferson, who throughout a life dedicated to public service was forever fending off the siren's call of mathematics, knew of what he spoke.

From time to time the aging Jefferson liked to complain about his declining mathematical powers. "After 40. years of abstraction from [astronomy]," he wrote Dr. Robert Patterson in 1811, "and my mathematical acquirements coated over with rust, I find myself equal only to such simple operations & practices in it as serve to amuse me."[73] He repeated the same complaint some months later to Bishop James Madison (not the president), telling him, "I have been for some time rubbing up my Mathematics from the rust contracted by 50. years pursuits of a different kind." Nevertheless, he continued on a more optimistic note: "Thanks to the good foundation laid at College by my old master & friend Small, I am doing it with a delight & success beyond my expectation."[74]

Rusty or not, Jefferson remained engaged with mathematics to his dying day. His library, which he sold to Congress in 1815 after the British burned down the original collection in the War of 1812, underscores his passion: it contains the works of most of the leading mathematicians of the Enlightenment, from Newton to Lagrange, as well as works by Guillaume de L'Hôpital, Maclaurin, Daniel Bernoulli, Euler, Jean Etienne Montucla, and John Playfair.[75] One would be hard-pressed to find a comparable collection of mathematical works anywhere in the world at that time. Forever interested in mathematical education, Jefferson insisted that the newly formed United States Military Academy at West Point would instruct the young cadets in the latest mathematical techniques. With the École Polytechnique as a model of elite mathematical training, and with graduates of that celebrated Parisian school serving on the faculty, West Point soon became the leading center of mathematical instruction in the United States.[76] Yet as if to emphasize that the value of mathematics was not limited to its use in warfare, Jefferson also insisted that it take pride of place in the liberal arts curriculum of the University of Virginia. Thanks to Jefferson's guidelines, the young university in the bucolic town of Charlottesville appointed a succession of brilliant mathematics professors, some of them lured from Europe. The university's most celebrated appointee was J. J. Sylvester, one of the geniuses of nineteenth-century mathematics, who served on the faculty in 1842.[77]

Had history taken a different turn, we may well be celebrating Jefferson today as the first great American mathematician rather than as a Founding Father of the United States. To be sure, Jefferson never made any original contributions to mathematics, and there is no way of knowing whether his considerable mathematical gifts would have been sufficient to earn him a place among the great geometers of his age. Yet the high regard in which he was held by his mathematical mentor William Small, as well as his lifelong devotion to the field in spite of very unfavorable circumstances, suggests that he may well have earned himself a reputation in the field. And even as his life and historical winds carried him in a different direction, he remained an avid amateur mathematician who always kept abreast of the latest developments in the field and looked at the world through mathematical eyes. It is hardly surprising then that his contemporaries, fully aware of his mathematical bent and talents, were happy to let him take the lead on issues that seemed to demand mathematical acumen.

Making a Mathematical Republic

In June 1783, shortly after the end of the Revolutionary War, Jefferson was elected to the Continental Congress as a member of the Virginia delegation, and he immediately plunged himself into the most mathematically challenging issues confronting the young republic. The first to come his way, and the one that had the greatest urgency for daily life, was establishing a currency for the Union. Under British rule, the official coin of the colonies was the English pound, divided into twenty shillings, each of which was worth twelve pennies. This, however, was largely a fiction, as the colonies were flooded by a vast array of European coins whose exchange rates were murky at best: half joes, doubloons, pistoles, pistareens, and rix-dollars all circulated alongside the English pound. A longtime favorite was the Dutch dog dollar, named after the canine appearance of its lion emblem, but by the 1780s the most popular coin circulating on the Atlantic Seaboard was the Spanish dollar, also known as a "piece of eight" for its standard division into reales. But the value of even this shared currency fluctuated wildly among the states, ranging from five shillings in Georgia to thirty-two and a half shillings across the border in South Carolina.[78] To manage daily life in this chaotic state of affairs required of each citizen an uncommon level of financial wizardry and inevitably opened the way to endemic suspicions of fraud, whether real or imagined.

By the time Jefferson got involved in devising a new currency, Congress's superintendent of finance, Robert Morris, and his assistant, the unrelated Gouverneur Morris, had already come up with a plan. In the Morrises' proposal, the basic currency unit was tiny—a mere 1/1140 of a Spanish dollar. The advantage of such a tiny base, they argued, was that it would be possible to convert it to each and every one of the different coins then in use. A fixed number of these units would equal every coin in circulation, making it easy for merchants to convert their old account books to the new currency.[79]

Jefferson, however, liked neither the plan nor its authors. The rotund Pennsylvanian, Robert Morris, was the leading financial speculator in the colonies before the revolution and in the United States after it. A signer of the Declaration of Independence, he had put his capital and his financial talents in the service of the shared struggle, earning him the sobriquet of "financier of the revolution" and the gratitude of George Washington. Winning over Jefferson, who disliked cities and despised financiers, was a different matter. In Robert Morris he recognized the leader of what he called "that Speculating phalanx," which was out to exploit the people and the Union for their own financial gain and destroy his dream of a virtuous agrarian America.[80] His assistant was little better. Born to an old and respected family, young Gouverneur was a notorious rake and bon vivant of the bustling city of New York. Though not as rich as the older Morris, he was clearly on his way to joining him among the American speculating elite.

As for the plan itself, Jefferson considered the basic unit far too small and entirely arbitrary. True, it would facilitate to some extent the bookkeeping practices of merchants, who dealt with many different circulating coins and would like a simple way to convert them to a standard American currency. At the same time, such a tiny unit would make the daily transactions of pretty much everybody else unreasonably cumbersome. Every simple purchase of a sack of flour or a farm animal would involve complex calculations with five, six, or seven figures, which would not only waste time but also hugely increase the chance of error and the opportunity for fraud. Jefferson, who considered merchants and financiers a scourge, cared nothing for their convenience. But he cared very much for the common people, and in particular for farmers, whom he considered the guardians of republican virtue. To him, a currency system that accommodated wealthy merchants at the expense of common farmers had its priorities inverted and should be rejected outright. On top of this, Jefferson had good reason to suspect that Morris was angling to run the US mint that would produce his new coins, thereby taking control of the nation's money supply. Handing such a vital national resource to the republic's wealthiest financier was to Jefferson a nightmarish proposition that must never be allowed to pass.[81]

To prevent such a calamity, Jefferson in early 1784 presented Congress with his own currency plan. Instead of inventing a new unit, as the Morrises had done, he proposed relying on the most familiar coin then in use, the Spanish dollar, which would henceforth be called the US dollar.[82] And while this seemed to be a move to preserve the status quo, Jefferson's innovation lay in the manner in which he proposed to divide new currency. Like practically all coins of the era, the Spanish dollar was commonly divided into halves, which were in turn halved, halved, and halved again, leading to half-dollars, quarter-dollars, eighth-dollars (or "reales"), and so on. Jefferson, however, proposed to divide the US dollar into tenths, hundredths, and thousandths, arguing that such divisions are mathematically simpler and would lead to fewer mistakes. This seems self-evident to us today but was far less so to eighteenth-century Americans who since childhood had been used to repeated divisions by two. Nevertheless, Jefferson's proposal carried the day, perhaps more because of concern about the Morrises' intentions than because of the persuasiveness of the Virginian's argument. On July 6, 1785, Congress decreed that "the money unit of the United States of America shall be one dollar" and that "the several pieces shall increase in decimal ratio."[83] It would be several more years and numerous congressional committees and recommendations before a mint was established and the first US dollar went into circulation, but the resolution of 1785 proved decisive. Thanks to Jefferson, the US dollar became the first decimal currency in the world, serving as a model for all modern currencies. And thanks to the enduring power of old habits, the Spanish dollar nevertheless had its subtle revenge: the US dollar is among the very few currencies in the world whose most common division, to this day, is into quarters.

The second issue to which Jefferson devoted his mathematical talents in the 1780s—and which would continue to occupy him for another decade—was the reform of weights and measures, which were in as chaotic a state as the currency at the time. To us, the two issues are entirely unrelated: weights and measures refer to actual physical objects and territories in the world, whereas the value of a coin is an arbitrary convention, which people agree upon for their convenience. This seems so natural that we sometimes forget just how novel and recent this notion of currency is. Because in fact, from the earliest uses of coinage in the ancient Mediterranean right up until the United States fully abandoned the gold standard in 1971, the value of a currency was ultimately dependent on the weight of precious metals. For Jefferson, as for his contemporaries, setting a clear value and standard divisions for coinage was inseparable from doing precisely the same for units of weight, as well as the related ones of length, area, and volume. And so the first paper in Jefferson's hand that indicates his thoughts on the matter, dating from the spring of 1784, is entitled, tellingly, "Notes on Coinage."[84]

Ever the Enlightenment universalist, Jefferson was determined to establish his units of measurement on the natural structure of the Earth. This, to be sure, was not an entirely novel idea. Back in 1671, in his book *Mesure de la Terre*, the French astronomer and cartographer Jean Piccard (1620–82) suggested basing a system of measurements on the circumference of the Earth. Piccard had used triangulation to measure the length of a degree of latitude with unprecedented accuracy. By multiplying his result by 360, he arrived at a figure for the circumference of the Earth that is less than half a percent smaller than the true figure. His results were later improved upon by four successive generations of astronomers named Cassini, all of whom served the French kings as directors of the Paris Observatory, as well as by measurements made on expeditions to Lapland and South America in the 1730s, sponsored by the Paris Academy of Sciences. On the basis of their accumulated results, Jefferson calculated that the length of one minute (i.e., one-sixtieth of a degree of latitude) is 6,086.4 feet, a unit that he referred to as the "geographical mile" and, more optimistically, the "American mile." He then proposed to redefine traditional measures as decimal fractions of the new unit: A "furlong" would be 608.64 feet, a "chain" 60.864 feet, and a "pace" 6.0864 feet.[85]

This was as far as things went in 1784, though Jefferson never abandoned his hopes of reforming American measures and setting them as a rational example to the world. During the five years he spent in Paris, from 1784 to 1789, he consorted with French scientists who were working on their own reform of weights and measures, and in particular with his friend the Marquis de Condorcet (1743–94), permanent secretary of the Paris Academy. The Earth, the French savants argued, was not a true sphere but a very imperfect one, beset by irregular bulges and depressions. This means not only that the precise circumference is extremely difficult to measure but also that it is very difficult to define: it varies considerably depending on which great circle is measured. Consequently, the dimension of the Earth is not a reliable basis for a universal system of measurement.

Condorcet had a better idea: instead of using a fraction of the circumference of the Earth as a basic unit of measurement, he proposed using the length of the "second's pendulum"—the length of a pendulum that completes a full cycle every second. The standard is based on Galileo's observation that a pendulum's period of oscillation is dependent on its length alone. Newton had proved it and calculated that the length of a pendulum that completes a cycle in exactly one second is 39.2 inches. Jefferson was immediately converted: since the length of the pendulum could be measured directly anywhere in the world, this seemed to Jefferson an ideal basis for a natural and universal system of measurements.

In 1790, back from France and serving as Secretary of State, Jefferson believed the time for reform had finally come. On July 13, he submitted to Congress a detailed "Plan for Establishing Uniformity in the Coinage, Weights and Measures of the United States," which was based on the length of a second's pendulum at a latitude of forty-five degrees (thereby slightly shorter than Newton's number for London's latitude of fifty-one degrees). Since a perfect pendulum—in which the string is weightless and all weight is concentrated at a single point—is impossible to produce, Jefferson replaced it with a pendulum made of a uniform swinging rod of 58.8 inches. One-fifth of this, or slightly less than 12 traditional inches, he called a "foot," which served as the basic unit of his system. One cubic foot would be designated the new "bushel," the basic unit of volume, and a bushel of water would weigh a thousand new "ounces." And the weight of the dollar was to be adjusted so that it came to be exactly one ounce of silver.[86]

Jefferson's proposal of 1790 was a perfect Enlightenment reform plan. Derived from nature itself, it was based not on human tradition but on the deep order of the universe. It was rational and systematic, replacing all the arbitrary, variable, and confusing units then in use with a single universally applicable system. It was comprehensive, tying together measures of length, volume, weight, and coinage in a single interconnected rational system, and it was decimal, making it easy to convert the various units into one another. Such a simple and rational system would wrench the units of measurement from the hands of the aristocratic elites, who had used them to exploit their underlings, and make them accessible to common people everywhere. In its ambition it captured Jefferson's desire to start anew, demolish an old world built on irrational tradition and superstition, and replace it with one founded on the clear light of reason.

If Jefferson's proposal had passed in Congress, then the United States today would likely be known as the birthplace of the metric system rather than as one of the last (and by far the most significant) nonmetric holdouts. But Jefferson held back: reports from revolutionary Paris indicated that Jefferson's old friends, including Condorcet, Lagrange, and the chemist Antoine Lavoisier had been appointed to a committee for the reform of weights and measures. Confident that the recommendations of the committee would closely parallel his own, Jefferson asked Congress not to vote on his proposal but to wait to hear the details of the French report. Much better, he reasoned, to tweak his system to correspond with the Parisian one than to end up with two competing systems, each claiming to be "universal."

But when the committee's report finally arrived in America, Jefferson was stunned. His French friends, who a few years before had convinced him to abandon the circumference of the Earth as the basis of measurement in favor of the second's pendulum, had inexplicably reversed course. In their report, published on March 19, 1791, they discuss the advantages of the second's pendulum at considerable length, praising its certainty and accuracy. Then, with little warning, they add: "However we ought to observe that the unit thus derived contains something arbitrary"—the second itself. A second, they explain, is the 86,400th part of the length of a day, which is an arbitrary division and therefore not sufficiently "natural." The fact that any unit of measurement based on the meridian would require an equally arbitrary division of a natural magnitude—the circumference of the Earth— troubled them far less for some reason. And so they abandoned the pendulum and instead set about once again to determine the precise length of a meridian, dispatching astronomers Pierre Méchain and Jean-Baptiste Delambre to triangulate the meridian between Dunkirk and Barcelona.[87]

The French savants' sudden abandonment of the pendulum standard pulled the rug from under Jefferson's proposal. The supposedly universal proposal that would set the standard for the entire world turned out to be nothing of the sort. It now appeared to be just an overly ambitious reform effort that would disrupt the work and the daily habits of all Americans and wreak havoc on commercial transactions, all to no obvious gain. And so, much to his chagrin, Jefferson's reform of weights and measures was never adopted by Congress, leaving the country with a dizzying array of incompatible units for years to come.

In taking on his successful reform of the currency and his unsuccessful reform of weights and measures, Jefferson was certainly demonstrating impressive mathematical acumen. Few if any of his fellow legislators, with the possible exception of financiers like the

Morrises, would have been capable of producing such systematic and mathematically coherent consistent proposals. Yet the ways in which he applied his mathematical talents are even more significant. The power of mathematics to reform human institutions, for Jefferson, lay in the fact that it demolishes and ultimately replaces the layers upon layers of irrational human traditions. Mathematical order exists in a space uncluttered by a plethora of different national currencies or an overabundance of units of measurement. It replaces it with a perfect, uniform, and undifferentiated mathematical space where perfect rational order prevails. It is in such a space, free of the hindrances and obstacles accumulated by detritus of countless generations, that men can finally be free.

In the domains of currency and weights and measures, such spaces are metaphorical, pointing to the power of mathematical order to free men from the shackles of tradition. But in the American West, these spaces were entirely real. For here was the true perfect, uniform, and unlimited space, with no past and no traditions, where enterprising men would be free to build a new nation. It was for Jefferson the physical incarnation of absolute mathematical space, with its promise of a new rational order of equality and freedom. He set about to mark it as such.

A Mathematical Empire

Seven Ranges

On the morning of September 20, 1785, surveyor Thomas Hutchins stood in the middle of an ancient forest on the north bank of the Ohio River. The scene was enchanting: "The whole of the above distance," he wrote, "is shaded with black and white Walnut trees," as well as with "Black, Red, and an abundance of White Oaks, some Cherry Tree, Elm, Hoop-Ash, and great quantities of Hickory, Sarsarfrax, Dogwood, and innumerable and uncommonly large Grape Vines."[1] It was a lovely landscape, one that has since been forever changed, first by the land clearing of agricultural settlers, then by the urban and industrial growth of eastern Ohio. But Hutchins was not there to enjoy the view. He had a job to do, for he had been charged by Congress with launching the rectangular survey of all western lands. The point where he was standing, west of the border of Pennsylvania where it crosses the Ohio River, was the point where the survey would begin.[2]

It was the highest appointment any surveyor could hope for in America, if not the world, and Hutchins came by it honorably. Born in New Jersey in 1730, he was orphaned as a teenager, and rather than rely on his family for help, he went west and obtained an appointment as an ensign in the British army. During the French and Indian War, he was commissioned an officer in the Royal American Regiment, where he specialized in drawing maps and designing military fortifications and encampments. In 1763, shortly after peace was signed, he was part of a force sent from Philadelphia to relieve Fort Pitt (later Pittsburgh), which had been besieged by Indian tribes in what became known as Pontiac's War. In a bloody and brutal campaign, Hutchins distinguished himself as a brave and efficient soldier as well as a master engineer, and was charged with rebuilding the defenses of the damaged fort. In the following years, Hutchins was posted all across the West, from Louisiana to Florida, gaining a reputation as an expert frontiersman and the leading cartographer in British America.

Hutchins was a commissioned officer in the army of George III, an honor never granted to his contemporary, George Washington. But as the rift deepened between the American colonists and the king, Hutchins had no doubt where his loyalties lay. He fully identified with the colonists' hunger for western lands, and in his writings he described Louisiana as a rich land of plenty awaiting its settlers. "If we want it, I warrant it will be ours," he wrote enticingly. And even as the king issued his decree banning settlement west of the Appalachians, Hutchins contributed to his regiment's war memoirs an appendix on the construction of frontier settlements. Hutchins recommended that they should consist of a one-mile square of agricultural land surrounding a smaller defensive hub along a riverbank. This affinity for squares may well have recommended Hutchins to Congress when it came time to decide on the proper leader for the rectilinear survey.[3]

The outbreak of hostilities between the colonists and the British army found Hutchins in London, where he was trying to have his memoirs and maps printed and published. Knowing his value as a frontiersman and cartographer, his British commanders fully expected him to rejoin his regiment and fight for the king, even offering him a promotion. Yet Hutchins refused, and was ultimately arrested by the British, losing his commission, his maps, and his money in the process. With help from Benjamin Franklin, he made it back to America in time for the final stages of the war in the southern states. In appreciation of his loyalty and his contributions to the war effort, he was named "Geographer of the Southern Army" by an act of Congress. A few months later, Hutchins, along with Simeon De Witt, who served as "Geographer to the Main Army," was awarded the title of "Geographer to the United States of America." They were the only men ever to hold the office.[4]

And so it was that when Congress went looking for the man to conduct the grand survey of western lands, Hutchins was an obvious choice. A veteran frontiersman, no one could doubt his ability to survive, work, and thrive in the western wilderness. He was also an outstanding surveyor and cartographer who knew how to make use of the most advanced instruments, take all the necessary astronomical measurements, and turn them into a precise map of the region. And he was, as he had proved in a long and eventful career, an American patriot and a true believer in western expansion. Finally, after the resignation of his colleague De Witt in 1784, Hutchins was the only official "Geographer to the United States." As a result, the choice of Hutchins to launch the great survey of western land was not a hard one. In fact, it was well-nigh inevitable.

By the time the rectangular survey was completed a century and a half later, it would cover 1.4 billion acres in thirty states and stretch across North America, from the Appalachian Mountains to the Pacific Ocean and Alaska.[5] And it is fair to say that even in 1785 there were some in America, including Thomas Jefferson and possibly even Hutchins himself, who in their mind already envisioned the survey covering the entire continent. Yet for the moment Hutchins was assigned a far more modest, and seemingly achievable, goal. He was to plot a precise straight line westward from the point of beginning, divided into six-mile-long segments. The base line is known as "the Geographer's Line" in Hutchins's honor, and each six-mile segment along it is designated a "range," which is why Hutchins's survey area is known to this day as "the Seven Ranges." From the endpoint of each range, he was to plot a line heading directly southward, until it reached the Ohio River, and since the river flows roughly to the southwest, these lines would get longer as he moved westward. Each of the seven north–south lines would itself then be divided into six-mile segments, though the number of segments would vary from line to line depending on its length. Finally, straight east–west lines would be plotted at regular intervals south of the Geographer's Line and parallel to it, so that they would cross the north–south lines every six miles. As a result, the irregularly shaped area northwest of the Ohio River would be covered by a grid of precise and equal squares, six miles to each side.[6]

It was a challenging project, but nothing about it seemed beyond the capabilities of an experienced surveyor like Hutchins. To assist him, he had a team of around thirty men with varying degrees of expertise. Congress, fully aware of the historical significance of the survey, had decreed that Hutchins would be joined by thirteen surveyors, each from a different state of the Union, but as it happened, only eight bothered to make the trek into the western wilderness. The rest of the team were muscular axmen, who would clear their way of brush and trees, and chainmen, who would lay down the surveyor's chain time and time again along the line. On September 20, 1785, they began their work.

If by some strange time warp we were fortunate enough to observe an eighteenth-century surveying team, we would see them moving along a straight line like a caterpillar crawling along a tree branch. From the starting point, the surveyor, using a compass, would find a distant landmark in the direction to be surveyed, in this case precisely westward. A team of axmen would then be sent forward to clear away trees and brush and open a straight path toward the landmark. Through this path, the chainmen would follow: the rear man would hold down one end of the "Gunther chain" at the point of beginning, and the front man would then walk forward, unfurling the chain as he went until it was stretched to the limit, and mark the spot with a tally peg. The rear man would then move up to the tally peg, and the process would repeat itself time and time again. If, looking from the point of beginning, the rear man always covered the front man, then the cumulative line of all the sections together would be straight. And so Hutchins and his team proceeded, chain by chain, westward into the Allegheny Plateau, clearing a straight path before them. Each Gunther chain was twenty-two yards long, 80 chains made a mile, and 480 chains made a six-mile "range," where a new line, pointing southward, would begin.[7]

As it happens, they did not make it far. On October 8, just over two weeks after they set out, troubling news reached the Geographer's ear: a settler's trading post in the Delaware village of Tuscarawas, about fifty miles to the west, had been raided, supplies taken, and one of the traders killed. Hutchins was not surprised: the resident Delaware, Ottawa, Miami, and Shawnee Indians had been watching the surveying work with increasing alarm, knowing full well that the small band of surveyors would shortly be followed by a flood of new settlers. Whether the raid in Tuscarawas was intended as a message to his team or not, Hutchins took it as such. Feeling alone and exposed in Indian Country, he packed up his tools, rounded up his men, and headed back to the safety of Pittsburgh. The first season of the great rectangular survey had netted only a single line, less than four miles long. And since Congress was paying them two dollars for every mile surveyed, the surveyors could not have been pleased.[8]

Disappointed but not ready to surrender, Hutchins returned in 1786 for another season of surveying. With a nearly full complement of surveyors from the states and a promise of protection from the army, he confidently predicted he would complete the survey of thirteen ranges. He himself would extend the Geographer's Line farther west, and each of the state surveyors would be responsible for a southward line at the end of each range. The results, though somewhat improved, were still disappointing. Dogged again by rumors of an impending Indian attack, and having lost many of their horses to Indian raiders, Hutchins's team proceeded slowly and intermittently. By the end of the season, they had completed the survey of only four ranges, though in this case they had also included the rectilinear southward lines and the six-mile squares known as "townships." Congress, frustrated by Hutchins's snaillike progress and eager to replenish its coffers from the sale of land to settlers, intervened: it limited the survey to a mere seven ranges. These were completed in 1787, though without the presence of Hutchins, who had asked for leave to pursue other business. The first plots went on sale in New York City in September of that year.

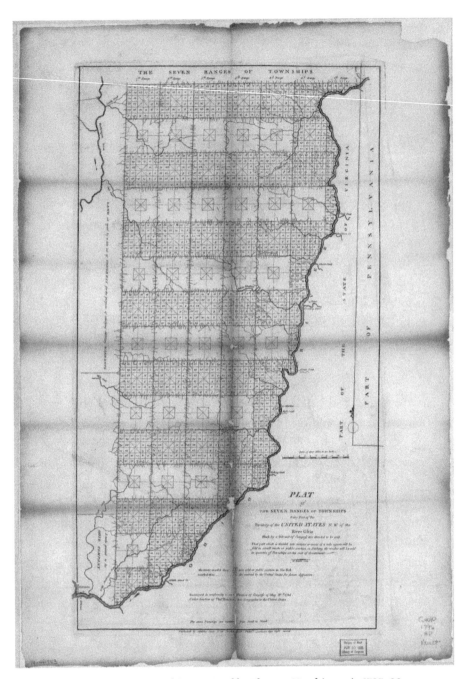

The Seven Ranges in Ohio, surveyed by Thomas Hutchinson in 1785–86.
Courtesy of the Library of Congress, Geography and Map Division.

Hutchins's survey of the Seven Ranges may not have been a complete failure, but it was certainly a disappointment. Despite the grand dreams of opening the West to settlement, only forty-two miles were marked on the east–west Geographer's Line and only ninety-one miles on the longest (and westernmost) north–south line.[9] And not only was the total area surveyed exceedingly modest but the quality of surveying was well below the standards of even the eighteenth century. Hutchins and his team seem to have proceeded west and south by relying on rough compass measurements, making few if any corrections along the way. As a result, the Geographer's Line dipped 1,500 feet south of its true course after forty-two miles, and the north–south lines were not straight and did not intersect the east–west lines in the right locations. The intersections between the sides of the townships were supposed to be at precise right angles but in practice were never so, leading to uneven "squares" that were shaped like diamonds or worse.

It is hard to conceive why a respected and competent surveyor such as Hutchins would produce such a shabby job, but it may simply be the fact that he was overwhelmed by the challenge, by his newfound prominence, and by the eyes of Congress upon him. The Geographer, after all, had made his name as a brave and self-sufficient frontiersman, an excellent independent cartographer and surveyor. But the task confronting him at the Seven Ranges was very different: here he was charged with managing a group of self-important and independent-minded state surveyors and rowdy bands of axmen and chainmen. Making all of them pull together in a common cause was a challenge of a kind Hutchins had never faced. And so, whereas Hutchins began the survey as the universally admired "Geographer to the United States," he ended it with his reputation much tarnished, and once more making his living by hiring out his skills to settle local boundary disputes. He died only two years after the completion of the Seven Ranges survey, at the age of fifty-eight.[10]

The survey of the Seven Ranges dealt a blow to Hutchins's reputation among his contemporaries. But it also secured his place in history. For despite its shortcomings—and there were many—the survey set in motion a process that irreversibly transformed the landscape of North America. Hutchins's survey recast the rugged landscape of the Allegheny Plateau, with its hills, ravines, and warlike Indians, as a uniform space defined by an abstract rectilinear grid. It would take another century and a half, but by the 1930s nearly all the territory of the continental United States west of the Appalachians would be similarly transformed. The woods and lakes near the Canadian border, the open prairies of the Midwest, and the arid deserts of the Southwest would all be divided into neat squares, just like the walnut, oak, and cherry forest traversed by Hutchins and his men. Following the template set at the Seven Ranges, the West would become a graph-paper-like terrain that could be sold, settled, and cultivated, square by square, by eager settlers.[11] It would be Jefferson's open, unlimited, and unfettered space, where free and enterprising men will found a new "Empire of Liberty."[12]

Jefferson's Plan

Jefferson's unshakable belief that the settlement of the West held the key to establishing a free society in America went back to his earliest days in public life. Already in 1774, in *A Summary View of the Rights of British America*, he had insisted that the colonists had a right to settle anywhere they wished in the vast expanse of the continent. There they would be free to set up whatever form of government they saw fit, unencumbered by king, aristocracy, or the traditions and hierarchies that were the bane European society. In the years that followed, as the conflict with the king became an open rebellion and a then long and bloody war, there was little Jefferson could do to advance his dream of western liberty. In the desperate struggle for survival of the young United States, neither he nor his contemporaries could expend time or energy on dreams of a glorious future beyond the Appalachian Mountains. But when the peace treaty was signed in 1783, Jefferson immediately sprang into action. He would make sure that the western lands, now abandoned by the British, would become the incubators of American democracy.

Before anything could be done, Jefferson recognized, the territories in question had to come under the jurisdiction of the government of the United States. As things stood at the end of the war, the western territories were a confused jumble of claims by individual states. Virginia, as the original colony, could point to its ancient charter and lay claim to territories from the Great Lakes to St. Louis, and Massachusetts's charter granted it lands "from the Atlantick . . . to the South Sea [i.e., the Pacific]."[13] New York lay claim to most of the territory of the modern states of Ohio, Michigan, Indiana, and Illinois, and other states simply extended their northern and southern boundaries westward, to the Mississippi and beyond.[14] As long as this was the case, Jefferson realized, the new settlers of the West would be dominated by the distant legislatures of the Eastern Seaboard just as the American colonies were oppressed by the distant rule of the London government. Instead of being a virgin land for men to forge a new nation, they would be a jumble of different and contradictory laws emanating from afar and ruling men's lives.

Jefferson was far from alone in arguing that the western territories belonged to the Union as a whole. Washington was just as insistent on this point, as were other revolutionary leaders who demanded that the states cede their claims to Congress.[15] One by one the states fell in line. The first to cede their western territories were also the states with the most extensive claims, New York and Virginia, and the others, willingly or reluctantly, followed suit. The last holdout was Georgia, which gave up its territories in 1802, a full twelve years after all other claims had been settled.

In 1784, Jefferson was merely a congressional delegate in Philadelphia, though as a former governor of Virginia and (though unbeknownst to many at the time) the author of the Declaration of Independence, an uncommonly prominent one. On March 1, he led his state delegation in ceding Virginia's territorial claims and immediately began pushing for legislation that would turn his dreams of the West into reality. The very same day, he submitted a report to Congress, written in his own hand, detailing how the territory would be administered.[16] The area between the Appalachians and the

Mississippi, Jefferson proposed, would be divided into states in the following manner: The northern and southern boundaries of each state would follow "parallels of latitude," with each state covering two degrees of latitude beginning at thirty-one degrees north and moving northward to the forty-seventh parallel. The eastern and western borders would follow regularly spaced meridians, except in the western edge of the territory, where the Mississippi River would serve as a natural boundary. The result would be a regular grid of sixteen new states. All would be roughly rectangular in shape except for the western border of the westernmost states, bounded by the river, and the northernmost states, which would fit into the irregularly shaped peninsulas of the Great Lakes. Jefferson even suggested names for the new states: the exotic-sounding and long-forgotten Sylvania, Cherronesus, Assenisipia, and Metropotamia, but also the more familiar Michigania and Illinoia, which survive little changed to this day.[17]

With the states' boundaries established, Jefferson turned to the people who would settle them. The vast majority, he assumed, would be white agricultural settlers who had purchased land from the government, the very people he described as "the chosen people of God ... whose breasts he had made his peculiar deposit for substantial and genuine virtue."[18] In the early stages, when the population was sparse, "their free males of full age" would "meet together for the purpose of establishing a temporary government" based on the constitution and law of one of the existing states. Once the settlers' population had reached twenty thousand, they would call a "Convention of representatives to establish a permanent Constitution & Government for themselves."[19] When this was done and the population of the new state had grown to the point that it equaled the least populous of the thirteen original states, "such state shall be admitted by it's delegates into the Congress of the United States, on an equal footing with the said original states."[20]

Jefferson was, in effect, trying to enact into law the very rights he had insisted on in the *Summary View* of 1774: the colonists' unfettered right to settle where they chose in the "empty" West and to establish a government of their choice. But to ensure that the West and its settlers would indeed become the backbone of the democratic republic, Jefferson added a few conditions. The new territories, he decreed, must remain part of the United States, and its settlers must be subject to its laws and obey its government. Just as significantly, the settlers, for all their unfettered freedom, cannot choose to place themselves under the rule of a king or an aristocracy. "Their respective Governments," he ruled, "shall be in republican forms, and shall admit no person to be a citizen, who holds any hereditary title." While Jefferson assumed that the colonists would naturally settle on republican forms of government as part of the United States, he was taking no chances. Only free citizens, democratically governed, could found the Empire of Liberty.

It did not take long for Congress to consider and act upon Jefferson's plan. On April 23, less than two months after Jefferson submitted his scheme, Congress voted on and passed a revised version of it. Commonly known simply as "the Ordinance of 1784," it kept most of Jefferson's grid intact, including all the evenly spaced latitude lines but one, to serve as boundaries between the states. The one line that was dropped was replaced by the Ohio River, which, flowing from east to west, served as a natural north–south boundary between states. For whatever reason, whether because the delegates were unimpressed with Jefferson's creative imagination or because they wanted to leave such matters in the hands of the future settlers, the Ordinance of 1784 included no mention of state names. Sylvania, Cherronesus, Assenisipia, and Metropotamia disappeared forever from the map of the United States.

By the time Congress passed the Ordinance of 1784, Jefferson had moved on to a question he considered even more fundamental than the formation of states: the form and shape of landholdings in the West. For as his proposal to Congress makes clear, the states themselves are arbitrary creations. They are legal and administrative entities, nothing more, and if their boundaries, or names, were altered or shuffled, nothing fundamental would change. Put differently, a settler in Michigania is no different from a settler in Illinoia, except for the fact that they happen to reside on opposite sides of an arbitrary line. The actual reality of settlement in the West had little to do with the made-up states; but it had everything to do with the plots of land settlers could acquire and cultivate, their size, their shape, and their relation to neighboring plots. And this Jefferson set out to define.

On March 2, 1784, only a day after submitting his "Report on Government for Western Territory" to Congress, Jefferson was appointed to head a new congressional committee on public lands. Less than two months later, he laid a proposal for an ordinance to govern the distribution and sale of lands in the West before Congress.[21] "Be it ordained by the United States in Congress assembled," Jefferson's report proclaimed, "that the territory ceded by Individual states to the United States . . . shall be disposed of in the following manner." What followed was, in sheer scale and ambition, the boldest plan ever proposed for land distribution anywhere in the world. The territory "shall be divided into hundreds of ten geographical miles squared, each mile containing 6086 feet and four-tenths of a foot, by lines to be run and marked due north and south, and others crossing these at right angles, the first of which lines, each way, shall be at ten miles distant from the corners of the state within which they shall be." With these initial units in place, the division would then continue on a smaller scale: "These hundreds shall be subdivided into lots of one mile square each . . . by marked lines running in like manner due North and South, and others crossing these at right angles."[22]

The units Jefferson is using here would have been as strange to his contemporaries as they are to us. His definition of the "geographical mile" would have been well-nigh incomprehensible, if we did not consider that at the very same time that Jefferson was proposing his land scheme, he was also working on a reform of weights and measures. A total of 6,086 feet and 4 inches it will be recalled, was his best estimate for the length of one minute (or one-sixtieth of a degree) of latitude and was the basic unit of his entire measurement scheme. The "hundred" was an ancient term with a history going back centuries, but in England it referred to a subdivision of a shire or county, not to a particular land area. Some of the American colonies, and later states, did use it as a measurement unit, but others did not, and its quantitative value varied widely. Jefferson is here for the first time assigning the "hundred" a specific value as ten of his geographical miles squared.[23]

But other than the unfamiliar units, the scheme was simple in the extreme. Straight parallel lines running from east to west would be plotted every ten miles throughout the entire western territory. Since a geographical mile is one-sixtieth of a degree of latitude, and each state in Jefferson's scheme would cover two degrees or 120 geographical miles, it follows that exactly twelve such parallel lines would fit within each state. Every ten miles, these parallels would be intersected at right angles by straight north–south lines, which would divide the land into identical squares each measuring a single "hundred" of ten geographical miles per side. The hundreds would then be divided more finely by running east–west and north–south parallels every geographical mile. Once the survey was complete, the entire territory of the West would be covered by a fine uniform grid, made up of precise squares of one geographical mile per side.[24] These would be the plots of land, Jefferson proposed, sold to the restless settlers who were poised to flood into western territories.[25]

It was a simple plan, and its audacity lay precisely in its uncompromising simplicity. For Jefferson's proposal makes no accommodations for the landscape, for mountains, hills, rivers, forests, or deserts. It makes no accommodations for the very different climates that prevail in different parts of the territory, the different types of soil, or the different crops that may be grown in different areas. It makes no accommodations for history, for the way the land may have been used by earlier white settlers, or for the traditional ways land had been measured and divided in Europe or in the colonies of the Eastern Seaboard. Most of all, it makes no accommodation for the fact that even as Jefferson was writing up his plan, the land was populated by native peoples who had lived on it since time immemorial and who had no use for his rectilinear plan.

Ignoring all conditions and constraints, both natural and human, Jefferson treated every point precisely the same as any other point on the land area. In his proposal, the West, for all its immense diversity and richness, becomes a uniform abstract mathematical space, defined through a uniform abstract mathematical grid. It was, in effect, a great Newtonian void, like the one he was so familiar with from his studies with his mentor William Small. And much as Newton's God created the world in the emptiness of absolute space, so will American settlers—hardworking, brave, and enterprising—create their own world in the great expanse of the West.

In conceiving of America as an empty Newtonian space, in which settlers can act freely and unconstrained, Jefferson likely derived inspiration from his philosophical hero, John Locke. It was Locke, after all, who (at Bishop Stillingfleet's urging) revised his view of space from a "full" Cartesian concept to an "empty" Newtonian one to allow both God and men freedom of action. It was also Locke who then projected this vision onto America, insisting that since the Indians did not cultivate the land, it should be treated as entirely empty and open to settlement. Finally, while Locke did not project the infinite Cartesian grid wholesale onto the American continent, he did propose a more limited rectilinear land distribution plan for America. Working as an aide to Lord Shaftesbury, and later as secretary to the Board of Trade, Locke in 1699 proposed that the colony of South Carolina be divided into squares of twelve thousand acres, aligned with the points of the compass. Like Jefferson a century later, Locke was following Newton's lead: an abstract mathematical grid was the best way, and perhaps the only way, to take possession of a vast and empty space.[26]

Yet it was Jefferson, not Locke, who brought the vision of Newtonian America to its logical conclusion and then proposed to Congress that it should be inscribed on the land. To appreciate just how radical this proposal was, it would be useful to compare it to the more traditional ways of carving out land for settlement, as practiced in the thirteen American colonies. In Virginia and its fellow southern colonies, settlers venturing into western territories were given near-unfettered freedom to mark out their lands. They would scour the land, pick out the parts with the best soil and most accessible water supply, and exclude less-promising areas. They would then delineate the boundaries of their land by clearly marking large trees, boulders, and mounds of earth and running lines between them, in a process called "metes and bounds" surveying. The result, in the southern colonies, was a patchwork of irregularly shaped farms that enclosed the best lands and excluded the rest. Much of the land between the farms, in this system, went unclaimed, and the cultivated land was widely dispersed. Jefferson, a native of Virginia and the son of a surveyor, was well familiar with this southern style of surveying. He considered it an example that should be avoided.[27]

A more regular, and to Jefferson's mind more rational, pattern of surveying emerged in New England. Not only was the climate there harsher, making it that much harder for individuals to go it alone, but the settlers arrived as pilgrims, in tightly knit religious groups. Land was accordingly granted by the crown not to individuals but to the group, usually in squares measuring six to ten miles per side. These were then divided up among the members in a grid of more or less regular rectangular plots that covered the entire square. In this manner the settlers of New England not only lived and worshipped in close cohesive congregations but also farmed the land in tightly packed plots, neighbor next to neighbor with no gaps between them.[28]

There is no question that Jefferson's vision of the gridded West was inspired, at least in part, by the practice of the New England colonies. Both plans were based on large squares that were then carved up into standard rectangular or square plots, resulting in a gridded pattern.[29] There were, however, important differences. For one thing, the squares in New England were oriented not to the points of the compass but to whatever direction seemed suitable to the settlers and the crown. Since preferences differed from one township to another, the result was a patchwork of squares oriented in different directions, not the uniform grid Jefferson envisioned for the West.[30] Another difference is that in New England the primary holder of the land rights was the collective, which then divided the land among individual families as it saw fit. Nothing could have been further from Jefferson's plans or ideals. For him, it was always the individual settlers, unhindered by tradition, creed, or authoritarian communities, who would build their homes in the western territories and create a new nation. It was therefore inevitable that in Jefferson's plan each plot would be acquired directly by individual settlers, who owed nothing to broader collectives, religious or otherwise. The man who declared that all men are endowed with an inalienable right to "Life, Liberty, and the Pursuit of Happiness" forever insisted on the full autonomy of individuals and their right to make their own way in the world. In this respect he was true not only to his Enlightenment ideals but also to his Virginia roots.

But the key difference between Jefferson's grid and New England squares was this: Jefferson's plan transformed the space of a continent, recasting it as an abstract uniform mathematical space. The New England practice did nothing of the sort. In New England, a community of colonists arrived at a favorable location and set about dividing the land among themselves. They marked out their overall square and then divided it into smaller ones equitably and efficiently. But beyond these closed circumscribed boundaries, all remained as it was—a land of forests and rivers, where wild animals roamed freely and native peoples lived as ever before. The New England settlers cared nothing for these surrounding lands; as long as they were safe in their own civilized square township, the rest of the land could remain as wild and untamed as it had ever been.

Jefferson's plan for the West was different in both kind and degree. For Jefferson's ambition was not to create safe havens for settlers in a vast untamed land. Rather, he wished to transform the land itself, everywhere and forever, by turning it into a regular mathematical space in which his dream of an Empire for Liberty could be realized. Whereas the New England squares had boundaries, Jefferson's rectilinear grid had none; it would simply continue on and on until it encompassed the entire continent. And whereas the New England squares had a center, Jefferson's grid had none, as every point on the grid and every square and rectangle within it was the same as any other. In Jefferson's vision, the West was Newtonian absolute space, with no boundaries and no center, marked only by an austere mathematical grid. Neither natural wilderness nor the native human inhabitants had a place in Jefferson's vision for the West: at most, they were inconveniences to be overcome by the resources of the government and the enterprise of the settlers. For ultimately, the West was a blank slate, a vast and nearly unlimited emptiness in which the new American Adam will found a New World.[31]

The Measure of a New Land

Jefferson had high hopes for his proposal to carve up the western territories into regular squares. Each "hundred" in his scheme was made up of a hundred plots of one square geographical mile, and each of these would be occupied by a single settler family. These tillers of the land, equal in wealth and circumstances and living off the sweat of their brow, were for Jefferson the most virtuous of peoples, untainted by the endemic corruption of big cities. Together they will form the backbone of an egalitarian democratic society and ensure it will never fall prey to the pretensions and prerogatives of those of high birth or great wealth.[32]

But this grand experiment in social and political engineering was not the only thing Jefferson hoped to achieve with his gridding of the West: he also meant to write his new units of weight and measure into the geography of the United States. The key unit of measurement in Jefferson's proposal was the geographical mile, a unit he defined as the length of one minute of latitude, and which according to his calculations equaled 6,086 feet and 4 inches. It was an awkward unit that did not fit well with traditional measures then in use, but to Jefferson that mattered little. The important thing, for him, was that it was a natural unit derived from the dimensions of the earth and could therefore be used universally, independent of local traditions or the interests of the rich and powerful. Introducing such a unit into general use, however, presented a formidable challenge: not only were people everywhere used to conducting business using their own local units but their books and records going back decades were also written in those traditional measures, and introducing new ones would make them useless overnight and eventually well-nigh incomprehensible.

But Jefferson believed he knew how to overcome this understand-

Patterns of land division in Virginia (top), Vermont (bottom), and South Dakota (right), as seen from space. Contains modified Copernicus Sentinel data, 2022–2023, processed by Sentinel Hub.

able resistance to his geographical mile and the other measures derived from it. "In the scheme for disposing of soil," he wrote to his friend Francis Hopkinson a few days after submitting his proposal to Congress, "an happy opportunity offers of introducing into general use the geometrical mile, in such a manner that it cannot possibly fail of forcing its way on the people."[33] In other words, if the West were surveyed using the new unit and the grid and its squares measured by it, then it would become inseparable from the American landscape, and people would have no choice but to make use of it.

Jefferson was well aware of the challenges and opposition facing his scheme. "I doubt whether it can be carried through," he wrote with a touch of realism born of long experience. But "were it to prevail," he continued more hopefully, even more advantages would accrue. The regular ten-mile spacing between the grid lines, the revival of the "hundred" unit as the land area of each square, and its division into one hundred one-mile squares would train the people in another of Jefferson's favorite rational reforms: decimal reasoning. Such a reform, he continued, "would lay the foundation for a very dangerous proposition: that is, to subdivide this geometrical mile into 10 furlongs, each of these into 10 chains, each of these into 10 paces." The danger, he deadpanned, is "that should we introduce so heterodox a facility as the decimal arithmetic, we should all of us soon forget how to cypher." And what is true of the geographical mile is also true of the decimal dollar he was trying to establish throughout the states: "I have hopes that the same care to preserve an athletic strength of calculation will not permit us to lose the pound as a money unit, and its subdivision into 20ths, 240ths, and 960ths, as now generally practiced." Jefferson, remembered today mostly for his soaring rhetoric, was also capable of acerbic wit. "This is surely an age of innovation," he concluded with what can only be considered a touch of self-irony, "and America the focus of it!"[34]

What would have happened if Jefferson had stayed to shepherd his proposal through Congress and turn it into law, we will never know. Had he succeeded, and had the Great Western Grid been measured in the geographical mile, the furlong, the chain, and the pace, it is quite possible that we would still be using these measures today, making the United States a pioneer in decimal measurement, just as it was in decimal currency. But that very summer, as Congress was mulling over his proposal, Jefferson was sent to Paris, where he would soon become minister plenipotentiary to the court of Louis XVI. In his absence, the urgency with which Jefferson infused the discussions over the settlement of the West petered out, and the debates continued month after month. The proposal was debated, revised, submitted to committee, and debated again. Finally, on May 20, 1785, more than a year after Jefferson had submitted his original report, Congress passed a bill.

Jefferson's prediction that his proposal for the surveying and land distribution of the West would never get through Congress proved prophetic. With the Virginian far away, his cherished geographical mile and its subdivisions of furlong, chain, and pace were quickly disposed of, and with them any hope that the settlement of western lands would lead to a reform of weights and measures. In their place, a conservative-minded Congress chose to adhere to the traditional measures of surveying, based on the twenty-two-yard-long Gunther chain: ten chains made a (traditional) furlong, and eight furlongs made a mile. Also gone were Jefferson's "hundreds" and any other trace of his elegant decimal system. Instead of squares ten miles to a side, Congress opted for squares of six miles to a side, which it designated "townships." Each township was, in turn, divided into thirty-six lots of one square mile each.

Jefferson, meanwhile, was trying desperately to keep abreast of developments from faraway Paris. But in an age in which any transatlantic communication would take months to arrive, there was nothing he could do to affect the outcome. All that was left to him was to bombard his friends with requests for information, which was bound to be outdated by the time it arrived. James Monroe, who was doing his best to keep him informed, did not sugarcoat what Jefferson was sure to consider a butchery of his plan for the West: "It deviates I believe essentially from the one [of 1784]," he wrote bluntly to his fellow Virginian a few days before Congress voted on the revised bill.[35] By the time Jefferson read the letter, the deed was done.

And yet, despite Monroe's harsh verdict, the most important element of Jefferson's scheme remained intact: the grid itself. For just like Jefferson's original proposal, the bill of 1785 set about carving out the West into precise, regular, and uniform one-mile squares, oriented to the points of the compass. Just as in the Virginian's vision, the grid was not closed or localized but open ended and uncentered, and would continue on and on as far as surveyors were allowed to go or until the entire continent was marked over. If Congress had its way, the great American wilderness would become an abstract and uniform mathematical landscape, a living continent transformed into the endless monotony of graph paper.

The Battle for the Grid

It almost didn't happen. For when the division of the West into a checkerboard of rectangular parcels was first proposed, it generated a fierce and immediate opposition. Washington, for one, was adamantly opposed, writing to William Grayson (who had succeeded Jefferson at the head of the congressional committee) that "the lands are of so versatile a nature that to the end of time they will not, by those who are acquainted therewith, be purchased either in townships or by square miles."[36] This was perhaps to be expected from a southern landowner, who was used to the freeform practices of southern settlers. But an even more powerful protest came from Massachusetts, whose delegates twice proposed legislation that would have effectively eliminated the western grid. In May 1786, within a year of the passage of the ordinance, delegate Nathan Dane moved that "in dividing the said territory into townships due regard be had to the natural boundaries," rather than to "a rigid adherence to lines run east and west, north and south," warning that rectilinear inflexibility "would manifestly prejudice the sale and future condition of said townships."[37] Only a few days later, in what may have been a coordinated effort, Rufus King, one of the leaders of the Massachusetts delegation, proposed to eliminate the provision that all lines be run "by the true meridian" because this "will greatly delay

the survey of the said territory." This meant that the square townships did not have to be oriented to the points of the compass but
could point in any direction, as they did in New England.[38] King's
motion actually passed, leading Grayson to complain to Madison in
a letter that "the Surveyors are liberated from all kind of connection
with the stars."[39]

In the aftermath of Hutchins's disappointing survey of the Seven
Ranges, Jefferson's grid seemed doomed to oblivion. Surveys initiated
in those years in Kentucky and Tennessee, and even in Ohio itself,
did not use the grid pattern at all but fell back on the tried-and-true
Virginia method of "metes and bounds."[40] It is less than surprising
then that when Alexander Hamilton in 1790 submitted a report on
public lands to Congress, he ignored the rectilinear requirements of
the grid altogether and recommended accommodating local conditions on the ground.[41]

When the issue of public lands was once again brought before
Congress in 1796, opposition to Jefferson's grid flared up again, broad
and fierce as ever.[42] William Findley of Pennsylvania, for one, argued
for natural boundaries that would improve the "advantages of nature"
rather than ride roughshod over them;[43] John Nicholas of Virginia
argued sensibly that since "the country was not square, the lines
could not be run in squares";[44] and William Maclay of Pennsylvania
advocating demarking the plots by "rivers, creeks, etc.," and argued
that cardinal points might prove "the most inconvenient of any."[45]
Richard Spaight, the Congressional delegate and future governor of
North Carolina, no doubt spoke for many when he railed against "this
formal and hitherto unheard-of plan."[46] And yet, in the face of determined opposition, Congress in 1796 voted to affirm the grid plan of
1785, restoring the provision that it must be strictly aligned with the
points of the compass.

Today the great grid has become so much a part of the western
landscape, so self-evident, and so inescapable that we hardly ever give
it a thought. It therefore takes an effort to recapture just how radical
and unprecedented was the decision taken by Congress to impose
this regular mathematical pattern on the North American continent.

It went against tradition and experience, for nothing like it had ever been attempted before. Colonists had thrived and expanded for centuries in America and elsewhere without ever resorting to such a formal mathematical plan. It went against common sense since the land and climate differed enormously from one region to another and equal square plots would differ wildly in their value and productivity. It went against the landscape, for as Congressman Nicholas pointed out, "the land is not square," and hills, valleys, rivers, and creeks are incompatible with the abstract grid. And it went against the advice of the most authoritative and respected man in the colonies, George Washington himself, who was likely also among the most knowledgeable in matters of agriculture and planning.

And if that were not enough, there was still the question of whether the plan was even feasible. Was it even possible, given the surveying technology of the day, to overlay a single massive grid over a continent-sized landmass? As the experience of the Seven Ranges revealed, probably not. Over the course of three years, the most accomplished surveyor in the United States had managed to run lines only a few dozen miles in each direction, and even these deviated substantially from their intended course and intersected at odd angles. The idea that one could extend such a survey for thousands of miles and establish a consistent grid over the better part of a continent would seem to an impartial observer very unlikely, if not outright fantastical.

And yet, in the face of nature, tradition, authority, and feasibility, Congress voted to adopt Jefferson's vision and impose a vast mathematical grid on western lands. They did so not because it was practical, or easy, or self-evident—far from it, for the grid was a fantastical idea, and establishing it would be fantastically difficult, if not impossible. They did so, rather, because the vision of the grid carried with it a vision of the West as an empty open land awaiting its settlers. The congressional delegates who voted on the bills of 1785 and 1796 may not have entirely shared, or even fully understood, Jefferson's sweeping vision of an agricultural West, in which free and equal farmers will serve as the backbone of a free nation. But they shared and understood enough of it to follow his lead and create a western landscape in which his dream could become reality.

By the time Congress discussed and then passed the bill of 1785, the finances of the United States were in crisis. Many of the states, having borrowed heavily during the war, were on the verge of financial collapse, and the coffers of the young republic were nearly exhausted. This distressing state of affairs inserted a note of urgency into the deliberations over the administration of the western territories. The sale of vast tracts of land in the West, it was hoped, would provide the Union with a large and much-needed infusion of cash, and since this could no longer wait, Congress determined to begin the survey forthwith. Much of the text of the Ordinance of 1785 is, accordingly, a detailed plan for launching the western survey.

The survey, Congress decreed, would be conducted by thirteen surveyors, one from each state, under the supervision of the "Geographer." They "shall proceed to divide the said territory into townships of six miles square, by lines running due north and south, and others crossing these at right angles." They shall be paid at a rate of two dollars a mile each, "including the wages of chain carriers, markers, and every other expense attending the same." As for the point of beginning, "the first line, running north and south as aforesaid, shall begin on the river Ohio, at a point that shall be found to be due north from the western termination of a line, which has been run as the southern boundary of the state of Pennsylvania."[47] These were, needless to say, the instructions carried out only a few months later by Thomas Hutchins, Geographer to the United States. When he stood on the bank of the Ohio River, looking westward through an oak and walnut forest, Hutchins probably gave no thought to Jefferson or his dream of a free egalitarian West. It is much more likely that he was thinking of the daunting technical challenges that faced him and the band of unruly men he was commanding. Yet whether he was conscious of it or not, Hutchins was beginning the process that would imprint the austere mathematical landscape of Jefferson's imagination onto the rich, diverse, dynamic, and very real physical landscape of North America.

"A Hored Savage War"

The road from Hutchins's haphazard survey of the Seven Ranges to the ultimate gridding of America was long and tortuous, not at all like the elegant straight lines that crisscrossed America in Jefferson's grand vision. Already in 1786, as Hutchins's men were extending their chains through the Allegheny Plateau and periodically retreating to the safety of Pittsburgh, Congress was revising another key feature of Jefferson's plan: the role of states and their boundaries in the survey. In Jefferson's proposal of 1784, the states themselves served as giant squares with more or less the same number of "hundreds" within their border. Accordingly, the first east–west survey line would be the state's northern (or southern) border, and it would repeat every ten miles until, two degrees and twelve lines later, the southern (or northern) boundary would be reached.

The multiple states Jefferson was proposing, each stretching over two degrees of latitude, were roughly the same size as their brethren on the Atlantic Seaboard and therefore seemed to him of appropriate size. But his fellow Virginian, James Monroe, who was less inclined to Jeffersonian abstractions and more familiar with conditions on the ground, had his doubts. In January 1786, having returned from a trip to the Northwest Territory, he wrote to Jefferson in Paris that the land was not at all the bountiful Eden he had imagined. "A great part of the territory is miserably poor," the future president reported, and "consists of extensive plains which have not had from appearances and will not have a single bush on them for ages." Consequently, he continued, Jefferson's proposed states "will perhaps never contain a sufficient number of inhabitants to entitle them to membership in the confideracy."[48] The solution, he argued, was to gather the thinly scattered settlers in a few large states rather than divide them into many small ones whose numbers would be too small for self-governance or civic life.

Monroe's plan for large states, he insisted, was entirely in the interest of the settlers and would "render them substantial service." But he did not hide from Jefferson that there was another reason for his

dissatisfaction with the small-state plan: the interests of the westerners, he predicted, will "in many instances be opposed to ours," and if
they have many states, they will also have a correspondingly strong
voice in Congress. "Instead of weakening their [interests] and making it subservient to our purpose," he charged, "we have given it all
the possible strength we could." If Jefferson's plan was accepted, he
worried, the western states would soon come to rule the republic, at
the expense of the original thirteen states—and, most critically, of
Virginia.

But if Monroe thought that Jefferson would rally to the side of the
native state at the expense of the westerners, he had misjudged his
man. Jefferson, after all, was the man who believed that the settlers
of the West were the most virtuous of men—honest, hardworking,
and enterprising. It was they, and not the overly civilized citizens of
the Atlantic states, who will ensure the survival of the republic, and,
as far as he was concerned, the stronger their voice in government,
the better. Asking how to dispose of the western territories "so as to
produce the greatest . . . benefit to the inhabitants of the maritime
states," he wrote Monroe, was unconscionable. "It is a question which
good faith forbids us to receive into discussion" because it implies
treating the settlers as subjects rather than fellow citizens, an attitude that is not only unjust but will inevitably lead to rebellion and
secession. The proper question is how to dispose of the territories "so
as to produce the greatest degree of happiness to their inhabitants."
And to this, he believed, there could be only one answer: Americans,
he had no doubt, would much rather live in small states of around
30,000 square miles, "in which they can exist as a regular society,"
than be dispersed in giant states of 160,000 square miles, as Monroe
was proposing.[49]

Jefferson's moral authority was great, and his vision for the West
inspiring, but he was far away in Paris. Monroe, however, was with
the Congress in New York, and so it was his narrow pragmatism,
rather than Jefferson's broad idealism, that prevailed. Neither impressed nor cowed by his fellow Virginian's admonitions, Monroe
chaired a congressional committee that recommended dividing the

western territories into three to five large states rather than a dozen small states. As head of another committee, he then dealt a further blow to Jefferson's ideal of free settler communities by recommending that the territories not govern themselves but be administered by Congress-appointed governors and judges. Only when the population of a territory reached sixty thousand would it be eligible to become a self-governing state on a par with the original thirteen.

On July 13, 1787, Congress passed the Northwest Ordinance, which has governed the territorial expansion of the United States right through the addition of Alaska and Hawaii in 1959. According to the ordinance, and just as Monroe had proposed, all territories were administered by appointed officials until their populations were large enough to be admitted as states to the Union. Jefferson's prediction that such a tyrannical system would lead to armed resistance proved groundless, as the practical Monroe always knew it would. The ordinance also adopted Monroe's suggestion of dividing the West into large states, which explains a curious fact revealed by even a cursory glance at a map of the United States: the western states are invariably much larger than the original states of the Eastern Seaboard.[50]

Finally, as became clear in the following years when territories were surveyed and divided, Congress was far more willing than Jefferson to accommodate natural features at the expense of rigid squares. The borders of the states of Ohio, Indiana, Michigan, Wisconsin, and Illinois, which occupy the area known in the 1780s as the Northwest Territory, are marked for the most part by rivers and lakes, not by the straight lines and right angles Jefferson proposed. The result of all this is that Jefferson's plan of anchoring the great grid on the state borders fell by the wayside. Jefferson's Newtonian West, oriented to the points of the compass and marked by straight lines intersecting at right angles, survived, but it had become even more abstract than before. Now, it ignored not only the natural landscape and the native population but also the state lines set down by the government of the United States.

The first plots of land from the Seven Ranges went on sale in New York City between September 21 and October 9, 1787.[51] With Hutchins and his surveyors present and available to answer questions, twenty-seven six-mile square townships, fifteen of them divided into one-mile squares, were offered to prospective buyers. Hopes for the sale ran high, as a cash-starved Congress sought to capitalize on the greatest resource under its control—the huge territory between the Appalachian Mountains and the Mississippi River. The sale of these lands to settlers, it was hoped, would fill the public coffers, enhance the credit of the US government, and enable it to borrow funds from foreign lenders. As it turned out, however, while the expectations were great, the disappointment was even greater. In nearly three weeks, only one-third of the land, totaling around one hundred thousand acres, was sold, and half of that was to a single commercial speculator. Lands bordering the Ohio River generally found buyers, but few were willing to venture inland, farther away from that lifeline to civilization. When allowances are made for the cost of surveying and uncompleted purchases, the net income to the treasury from the land sales comes to a paltry $100,000.[52]

One need not look far to find the reasons for these disappointing results. For one thing, the price of $1 per acre demanded by the government was simply beyond the reach of many potential settlers, but since it was written into law, it could not be quickly adjusted when sales proved anemic. Add to that the shoddy quality of Hutchins's survey, which fairly guaranteed that boundaries and ownership would be disputed, and one gets a sense of why potential settlers were wary of handing over their hard-earned funds in exchange for ill-defined plots of land that they had never seen.[53]

The most significant reason for the settlers' reluctance, however, was the very real fear of attack on outposts deep inside Indian territory. Hutchins's own team, after all, made up of experienced and hardy frontiersmen, had been so alarmed by rumors of Indian raids that it repeatedly retreated to the safety of Pittsburgh. And if the US Army could not protect even an official government expedition, then it certainly could not be trusted to protect isolated private farms. What hope could a lonely settler have of long surviving in such a place? As long as one was close to the river, reinforcements in time of danger were possible and flight to civilization was always an option. But the farther inland settlers and their families ventured, the more likely they were to find themselves alone and unsupported, surrounded by Indians furious at their presence on their ancestral lands. It would be a brave soul indeed, or perhaps merely a rash one, who would spend what little money they had to buy land in the middle of the Seven Ranges.

With land sales disappointing and conditions on the ground unsettling, official surveying ground to a halt in 1788, with the completion of Hutchins's survey. Some surveying work continued for a while under the auspices of the private Ohio Company, which had purchased from the government a million acres adjacent to the Seven Ranges. Led by the indefatigable surveyor and Revolutionary War general Rufus Putnam (1738–1824), the company proceeded to measure and divide the land in the pattern set by Congress—a grid of six-mile square townships oriented to the points of the compass, with each township divided into one-mile square lots. To defend against potential attack, the company founded a fortified settlement on the northern bank of the Ohio River, which they named Marietta in honor of Queen Marie Antoinette of France. Putnam himself moved there with his family to supervise the work, bringing a touch of European elegance to the wild Ohio Valley.

Yet in the end, the private survey fared no better than the public one, and for many of the same reasons. Since the Ohio Company's territory was adjacent to the Seven Ranges, the rectilinear boundaries were supposed to continue uninterrupted from one to the other. But as Putnam's surveyors proved no more competent than Hutchins's men, the lines missed so badly that the boundary between the two surveys looked like a step ladder. The result was that instead of an elegant rectilinear grid, the land was marked with a patchwork of irregular and uneven plots of disputed boundaries—not a prospect likely to entice prospective buyers. More ominously, despite the company's military precautions and efforts to reach agreement with local tribes, the surveyors were constantly harassed by Indian war parties. Things got worse in the winter of 1790, when a column of troops was ambushed and defeated, and in the following year, when fourteen settlers were killed in the inland region of Muskingum.

"Our prospects are much changed," an alarmed Putnam wrote to Secretary of War Fisher Ames in January 1791. "Instead of peace and friendship with our Indian neighbors," he wrote in his typically eccentric script, "a hored Savage war Stairs us in the face the Indians seem determined on a general War."[54] In response to Putnam's pleas, President Washington, his erstwhile brother in arms, sent a punitive expedition to subdue the Indians under the command of General Arthur St. Clair, governor of the Ohio Territory. The result was even more disaster for Putnam and his settlers: on November 4, St. Clair's army was routed in battle with the Western Confederacy, a coalition of tribes led by the Miami chief Little Turtle and Shawnee chief Blue Jacket. With 630 men killed and another 270 wounded, as against about 50 Indians killed, it was the worst defeat ever suffered by the US Army in its Indian Wars. The Western Confederacy was victorious, and with no help in sight, many of the settlers packed up and headed back east. No more surveying would take place in Ohio in the foreseeable future.

The defeat of General St. Clair and the clearing of surveyors from the Ohio Valley was a severe blow to Jefferson and his vision for the West. The Indians, whom he was determined to ignore as much as possible, had inconveniently made their presence felt and put the brakes on his plan to grid the continent. And yet this unplanned hiatus in surveying also had a significant side benefit for Jefferson, who was by this time secretary of state in Washington's cabinet. The surveys of Hutchins and Putnam followed the rules laid out in the Ordinance of 1785, which relied on traditional measures and divided the territory into six-mile square townships. Every additional line and square surveyed using these dimensions would further entrench them as the standard American measurements, effectively putting an end to Jefferson's dream of introducing a decimal system of weights and measures. With the survey on hold for the time being, Jefferson and his allies saw an opening: if they could convince Congress to conduct the rest of the survey using

Jefferson's decimal units, then they could yet turn the tables on the traditionalists. By inscribing decimal measures on the western landscape, they would ensure that those would become the true American measures.[55]

Sensing an opportunity, Jefferson's allies in Congress swung into action. In April 1792, a three-man Senate committee of which Monroe was a member submitted its report. Using Jefferson's second's pendulum as its standard, it recommended that a second's rod be constructed and one-fifth of it be labeled a foot. The rest of the measurement was to be decimal: "The foot shall be divided into 10 inches, the inch into 10 lines, the line into 10 points; and that 10 feet shall make a decad, 10 decads a rood, 10 roods a furlong, and 10 furlongs a mile." Most critically, the committee recommended that these measures be used for the squares that would make up the surface of the United States.[56]

Unfortunately for Jefferson and Monroe, and perhaps for the young United States, the window of opportunity for a decimal America closed almost as soon as it had opened. First came the dispiriting news from Paris that the French savants gathered in the Committee on Weights and Measures had abandoned the second's pendulum and opted instead for one more measurement of the circumference of the Earth. Since the academicians ordered that the measurement take place along the meridian from Dunkirk to Barcelona, this would effectively make France the standard of the world—something that Englishmen, Germans, Spaniards, and others were not happy to concede. And while Jefferson was always friendly to France and its revolutionary regime, he nevertheless saw the abandonment of the pendulum in favor of the geographically determined "meter" as a betrayal of the universal ideals of the Enlightenment.

Then came the uplifting news that the situation in the West had improved—for surveyors and settlers, at least, if not for Indians. This was excellent for Jefferson's dream of surveying and settling the West, but it was terrible for his hope of writing his decimal units into the topography of the United States. It was a sad irony for Jefferson, who was a fervent American patriot, that the fortunes of his decimal reform rose and fell in inverse proportion to the military fortunes of his country.

What happened was this: ever since St. Clair's defeat in November 1791, a US Army force under General "Mad" Anthony Wayne had been working to rebuild the American military position west of the Alleghenies. Despite his nickname, and a reputation for recklessness earned in the Revolutionary War, Wayne proceeded cautiously. He built and manned a line of forts, trained his men in musketry and bayonet charge tactics, and marched his men up and down the territory while avoiding open battle with the victorious forces of the Western Confederacy. In the summer of 1794, he was finally ready to strike: after a brief campaign, he cornered the main Indian force on the banks of the Maumee River, near the modern-day city of Toledo. Hoping to slow the Americans' advance, Little Turtle and Blue Jacket positioned their forces in a forest clearing, behind trees recently felled by a storm. Unhindered, Wayne's finely trained troops swept forward in a bayonet charge, then chased down the retreating warriors with their cavalry. When a nearby English fort refused to open its gates to their Indian allies, the Indian route in the Battle of Fallen Timbers was complete. The following year, the leaders of the Western Confederacy signed the Treaty of Greenville, ceding their rights to all of Ohio and most of modern Indiana to the United States. With the Indians removed and the territories secured, the West was open for settlement.[57]

The Imperfect Grid

"The emigration to this country this fall surpasses that of any other season," proclaimed the *Pittsburgh Gazette* in October 1795, shortly after the Battle of Fallen Timbers. "We are informed," the report continued, "that the banks of the Monongahela are lined with people intent on the settlement on the Ohio and Kentucky."[58] The long-expected flood of settlers pouring over the Appalachians into the western plains was finally here, and Congress swung into action. It could no longer wait for the resolution of the interminable debates over the relative merits of the geographical mile versus the second's pendulum or the question of whether the transition to decimal measurements was worth the turmoil and disruption it would cause. Instead it needed a quick solution, to make sure that the sale of western lands would proceed in an orderly manner that would benefit the federal Treasury. On May 18, 1796, it passed a bill to do just that.

The "Act Providing for the Sale of the Lands of the United States," known simply as "the Land Act of 1796," ignored the elegant plans proposed by Jefferson and Monroe. The survey and sale of land, it decreed, shall proceed along the lines set in the act of 1785 that guided the survey of the Seven Ranges. As before, the land "shall be divided by north and south lines run according to the true meridian, and by others crossing them at right angles, so as to form townships of six miles square," which in turn would be divided into thirty-six "sections" of one square mile each, equal to 640 acres. "Whenever seven ranges of townships shall have been surveyed," the act stipulated, "the said sections of six hundred and forty acres . . . shall be offered for sale, at a public vendue."[59] Not wishing to rock the boat, and with the pressure from prospective settlers building, Congress opted in effect to extend the pattern set in place in the Seven Ranges over the entire area opened up by the Treaty of Greenville.

To supervise this grand undertaking, Congress decreed that "a Surveyor General shall be appointed." Washington wanted this person to be Simeon De Witt, a member of his wartime staff who—after Hutchins's death—was the sole surviving Geographer to the United States. De Witt, however, chose to continue as surveyor general for the State of New York, a position that would see him play a critical role in the gridding of Manhattan years later. The job therefore fell to the man who was already on the spot and already conducting a survey—Rufus Putnam. For the next seven years, Putnam energetically supervised the government survey from his home base in Marietta, much as he had the private survey of the Ohio Company some years before. He divided the territory into districts and assigned a different team of surveyors to each. In theory, since each team would survey the same lines along the same latitudes and longitudes, the lines would meet and form a uniform grid covering the entire territory. The reality, however, was very different, as Putnam the official general surveyor used the same methods as Putnam the private company surveyor.

The surveying teams had little scientific expertise and rarely bothered to check their position or direction with astronomical observations. They ran their lines using rule-of-thumb methods that were common in the profession and perfectly adequate for marking property boundaries—which was what surveyors most often were hired to do. But for surveying the vast northwestern territories, these methods were woefully inadequate. As a result, the lines run by one team rarely met with those run by another, local corrections had to be made at the intersection of the different districts, "square" plots were shaped like stairways or diamonds, and their orientation varied considerably from the points of the compass. In some places the same piece of land received different designations, depending on which team of surveyors was credited. In Jefferson's vision, the West would be transformed into an elegant mathematical grid of straight lines, right angles, and perfect squares. Putnam's survey produced anything but.

Putnam himself was not overly concerned with these shortcomings. A self-made man with scant formal education, he had little interest in Jefferson's abstract universal ideals and none at all in his vision of the West as a mathematical tabula rasa. The purpose of the survey, as far as he was concerned, was to sell land, and for that, it seemed to him, his survey methods were good enough. And indeed, Putnam's survey did allow a flood of people from the older states and from as far away as Europe to purchase land and settle in the West. But as would be the case repeatedly in the course of American western expansion, the aspiring yeomen farmers of Jefferson's dream were joined by a swarm of speculators happy to purchase vast tracts of western lands they had never seen and keep them off the market while their value went up and up. Purchasers with deep pockets could bribe government officials and judges to secure their claims or hire thugs to fend off rival claimants to the land. Sometimes they hired their own surveyors to remeasure the land in their own interest and then took their neighbors to court for trespassing. Whether things would have turned out better if Putnam's surveying methods had been more exacting is debatable. As it was, under Putnam's regime the flow of settlers to the West continued unabated. But so did discord, acrimony, speculation, and outright fraud.[60]

Throughout the presidency of John Adams (1735–1826), from 1797 to 1801, Putnam's tenure as surveyor general was secure. Popular with his underlings, universally admired in Ohio, and well-connected in Congress, he was well entrenched in his position, and it was clear that any effort to remove him would lead to a bitter political fight. Even more significantly, the sitting president, John Adams, was little interested in western settlement and expansion. Adams had grown up in the bustling port city of Boston and as a prominent lawyer had worked closely with the city's mercantile and shipping magnates. Unlike Jefferson, who envisioned a republic composed of small landowners spread across the continent, Adams believed that the survival and prosperity of the United States depended on commerce, a robust financial system, and most particularly on international trade. The vast expanses of the continent to the west, which to Jefferson were an endless source of fascination and promise, held no such interest to Adams. The future of the United States, he believed, lay in lucrative oceanic trade with Europe to the east, not in the endless dreary prairies to the west. As long as things were relatively peaceful in the west, and it did not breed malcontents and insurrectionists, he was happy to leave things as they were. Putnam, consequently, kept his job.

But on March 4, 1801, Thomas Jefferson was inaugurated president of the United States, and Putnam immediately found himself on shaky ground. The surveyor general had good reason to be concerned: He had never liked Jefferson, whom he considered disloyal to his patron, Washington, and in his correspondence he referred to Jefferson unabashedly as the "Arch Enemy." Jefferson was fully aware of the surveyor general's antipathy and reciprocated in kind. Furthermore, deeply engaged as he was in the settlement of the West, Jefferson was highly critical of Putnam's unscientific surveying approach and the constant stream of land disputes and lawsuits that stemmed from it. This in itself might have been grounds enough for Jefferson to replace the surveyor general, and if Putnam suspected his days in his post were numbered, he certainly had good reason.

For Jefferson, however, there was something far greater at stake than personal friction with his surveyor general or even the irregularities of land ownership. The survey of the West, for him, was not just about land sales but about the transformation of the West into a vast unresisting emptiness, with no center and no boundaries, a blank slate in which free and equal American settlers would establish an Empire for Liberty. It was precisely the boundless and monotonous mathematical grid that accomplished this, making the rich and diverse landscape of a continent into a formal mathematical space in which all places were alike and anything could be accomplished. For Jefferson, Putnam's shoddy survey, in which lines were crooked, directions skewed, and angles anything but rectilinear, didn't just create unnecessary legal and administrative problems; it subverted Jefferson's dream of recasting the West as a mathematically defined Newtonian void, waiting to be filled by enterprising farmers. Putnam had to go.

Though Jefferson, as president, may have had the power to replace his surveyor general at his pleasure, he did not do so immediately. Jefferson, after all, was the first leader of the opposition party to become president, and there was no established precedent as to what to do with the officeholders already in place and loyal to John Adams's Federalist regime. In the end Jefferson managed to replace nearly all public officials with loyal Republicans, thereby setting an example to all future administrations. In the process, however, he aroused fierce opposition and loud denunciations from his defeated opponents, who accused him of tyrannical rule.[61] Removing the popular and well-connected Putnam, who was also Washington's friend and brother in arms, was sure to cause an outcry. And so Jefferson chose to bide his time until his new administration was secure and an opportunity presented itself.

In the first two and a half years of his administration, Jefferson delegated his dealings with Putnam to Albert Gallatin, his able secretary of the Treasury. Gallatin communicated regularly with the surveyor general in a brisk, businesslike fashion that conveyed no warmth but also gave no warning of his impending demise. What finally tipped the balance may have been the Louisiana Purchase, which was signed in Paris on April 30, 1803, but news of which only reached Washington in July of that year. At a single stroke, nine hundred thousand square miles of territory were added to the four hundred thousand square miles already owned by the federal government. All this enormous territory needed to be surveyed, and if Jefferson had any hope of it being done with the mathematical precision he thought essential, Putnam could not be the one to carry it out.

And so in a letter dated September 21, 1803, without any preliminaries, Gallatin delivered the blow: "The President of the United States having appointed Jared Mansfield of Connecticut Surveyor General of the United States," the letter began, "I have to request that on his arrival at Marietta, you will deliver over to him the public papers, records, documents & other public property in your possession."[62] And so the deed was done. Gallatin added that Putnam "will be pleased to consider yourself authorized to act till Mr. Mansfield's arrival at Marietta, at which time your salary will cease & his will commence," but Putnam was neither pleased nor mollified. In his reply letter, he reported having done everything that Gallatin had asked, but added: "Perhaps you may imagine this conduct looks like passive obedience and nonresistance, or that I am courting favor; mistake me not; I have done no more than what I conceive to be the duty of any public officer in like circumstances."[63] In his memoirs he was even blunter about his dismissal: "It was done because I did not subscribe to the Measures of him whom I have called the Arch Enemy to Washington's Administration. Because I did not die or resign."[64]

Making the Mathematical West

Jared Mansfield (1759–1830), the man Jefferson picked as the new surveyor general, could not have been more different from his predecessor. Whereas Rufus Putnam was a self-made and self-educated frontiersman, Mansfield came from an old and respected New England family and had been educated at the best schools. Like several of his ancestors, he attended Yale College, though not without incident: in his senior year he was expelled for the theft of books from the library "and other discreditable escapades," though apparently he was eventually allowed to return and graduate. It is possible that it was the shock of his father's death a few months before that instigated this uncharacteristic act of youthful rebellion, though we may never know. Either way, it was the only blemish on a personal and professional life marked throughout by uncompromising rectitude.

In the decades after his graduation, and following a family tradition, Mansfield became rector and teacher at several elite grammar schools that prepared boys for entry to Harvard and Yale. It was in those years that he also discovered the true passion of his life: mathematics. In 1801 he published a collection of articles titled *Essays, Mathematical and Physical*, which covered topics ranging from negative quantities in algebra, analytic geometry, and Newtonian fluxions to gunnery and navigation. Both the scope and the originality of the work immediately established him as one of the foremost mathematicians in the United States. It also caught the eye of President Jefferson, who in 1802 appointed him to the faculty of the newly founded United States Military Academy at West Point. The following year, he made him surveyor general.[65]

If Rufus Putnam approached the job of surveyor as a frontiersman and land salesman, Mansfield approached it as a mathematician. And like all members of this select profession, from ancient times to our day, he would not tolerate sloppiness or inaccuracy. Mathematicians, after all, inhabit a world far more perfect than our own, one in which circles are perfectly circular, lines are perfectly straight, and the sum of the angles of all triangles is exactly—but exactly!—equal to two right angles. When called upon to conduct a survey, Mansfield worked to translate this ideal mathematical world onto the western landscape, by making sure that the lines he ran were indeed straight and the intersection between them exactly at right angles. And so, whereas his predecessor was content to rely on practical rules of thumb common among surveyors, Mansfield insisted on having at his disposal the most advanced and most accurate instruments then in existence. Following his appointment, he placed an order with London instrument makers, known to be the best in the world, for a "three-foot reflecting telescope, mounted in the best manner, Wollaston's *Catalogue of the Stars*, Maskelyne's *Observations and Tables*, a thirty inch Portable Transit Instrument and Therdolete, [and] An Astronomical Pendulum Clock."[66] These were the instruments that he took with him to the West and that ensured that his survey was incomparably more accurate and consistent than those of Hutchins and Putnam before him.

But even more basic than his reliance on advanced instruments was the way in which Mansfield conceived of his surveying task. Putnam, like Hutchins in his day, had focused on outlining the boundaries of plots, with a view for their quick sale. Mansfield's goal was quite different: it was to create an immaculate mathematical grid of straight lines and right angles that would encompass the entire land. The fact that the grid created square plots of land that were then sold to settlers was almost an afterthought, a by-product of the perfect grid. For Mansfield, after all, was a mathematician, and his task as he saw it was the mathematical gridding of the land. In that sense it is he, and not Hutchins or Putnam, who is the true originator of the Great American Grid.[67]

And so Mansfield went about the western survey precisely as a mathematician would outline a Cartesian grid on a blank piece of paper. First he ran a precise north–south line that he labeled a "Principal Meridian." Then he ran a precise east–west line westward from the principal meridian, which intersected it at a right angle at a point labeled "the point of beginning." This he labeled the "Base Line." The principal meridian and the base line were, in effect, the x and y Cartesian coordinates that define the geometrical space, and the point of beginning is, inevitably, the "origin" of the Cartesian coordinates. To complete the grid he then ran additional north–south lines, parallel to the principal meridian, every six miles along the base line and ran additional east–west lines, parallel to the base line, every six miles along the principal meridian. The result was a precise and regular grid of straight lines and right angles, dividing the land into pristine square plots, six miles to a side.

In a Cartesian coordinate system, as we have seen, every point on the plane is defined by two numbers, one from each of the coordinates. Exactly the same is true of Mansfield's grid. Every six-mile stretch along the base line was called a "range," and each range was numbered: the closest to the point of beginning was Range 1, the next one Range 2, and so on. Similarly, every six-mile stretch along the principal meridian was called a "township," and every township was numbered beginning with the one closest to the point of beginning: Township 1, Township 2, and so on. In this way, each six-mile square, no matter its location, is defined by a range number and a township number: "Range 1 West, Township 1 North," for example, is the square on the northwest (or, conventionally, upper left) of the point of beginning. The square defined as "Range 2 West, Township 3 North" is exactly six miles to the west and twelve miles to the north of it. Mansfield's grid is a perfect Cartesian plane.[68]

In principle, a single principal meridian and a single base line should have been sufficient to grid the entire continent. That, after all, is the case in mathematics, where a single x and a single y coordinate are sufficient to define every point on the infinite plane. In practice, however, even Mansfield's superior surveying methods were not up to the task of creating a continent-sized grid from a single set of coordinates. Inevitably, as a surveyor moves farther and farther away from the principal meridian and base line, small errors accumulate one on top of the other and turn into major errors that distort, and can ultimately dissolve, the grid. Mansfield's solution was to establish additional principal meridians throughout the territory, each with its own base line and each responsible for a different part of the vast overall grid.

And so Mansfield measured not only the First Principal Meridian, which today marks the border between Ohio and Indiana, but also the Second Principal Meridian, which runs through the center of Indiana, and the Third Principal Meridian, which runs through Illinois.

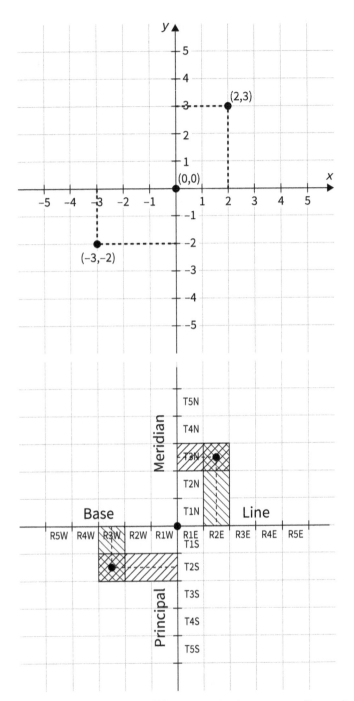

A Cartesian grid (top) and Mansfield's township/range system (bottom).

Over the following decades, as the survey moved westward, additional principal meridians were periodically run, some numbered, others named, all the way to the Willamette Meridian in the Oregon Territory and the Mount Diablo Meridian in California.[69] When Putnam in his day experimented with dividing his survey into smaller local ones, the result had been a disaster: the lines from different districts did not meet, the angles were not right, and the plots were not square. But with Mansfield and the surveyors who succeeded him, things were different. Thanks to his scrupulous measurements, the lines originating from different principal meridians and base lines almost always met as they were supposed to or intersected as they should—at right angles. The result was that despite the fact that the lines were measured from different points, there is only a single unified all-encompassing grid, stretching across the American West.

This still left open the problem of converging meridians. Since the north–south lines that form the borders of the townships follow the meridians, the curvature of the Earth dictates that the lines will draw closer to one another as one moves northward (the reverse is true in the Southern Hemisphere). This means that the northern side of each township will always be shorter than the southern side, and if townships are simply plotted one on top of the other, bounded by the same meridians, they will inevitably grow smaller and smaller. Mansfield had a solution: whenever the northern boundary of a township was at least sixty yards shorter than the southern boundary, he decreed, it was time to start over. Beginning with the principal meridian, new six-mile intervals would be measured along an east–west line designated a "standard parallel," to serve as the bases for new square townships. These corrections are almost invisible from the air, but woe to incautious motorists driving for mile after mile along roads running on the meridian between square townships to the east and west. If they forget the sudden correction that will unexpectedly interrupt the arrow-straight journey northward, they will end up in a field.[70]

Over time the grid grew more refined. In its original form the territory was to be divided into six-mile square townships, which were themselves divided into one-mile square sections to be sold to settlers.

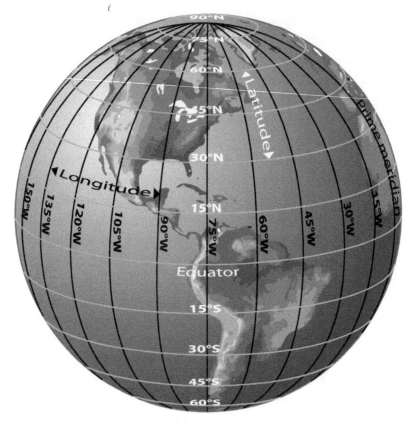

Whereas the lines of latitude remain parallel to one another,
the meridians, or lines of longitude, converge on the poles.

In practice, however, these square-mile (or 640-acre) units proved too expensive for individual settlers to buy and too large for them to cultivate on their own. As long as this was the case, there was danger, all too often realized, that most of the land would fall into the hands of deep-pocketed speculators, who would then rent it out to settlers. For Jefferson and his fellow Republicans, this was a nightmarish scenario, an inversion of their ideal of free private ownership of the land by the men who tilled it.

To prevent this, successive administrations kept reducing the minimum size of plots that could be sold to individual settlers, to bring it within their reach. In 1800 it was set at half a section (320 acres), in

1804 at a quarter section (160 acres), and in 1820 at a half-quarter section (80 acres). Finally, in 1832 the minimum size of a plot on which a farmer could bid was set at a quarter-quarter section, or 40 acres. This size was to remain the iconic plot of the American West, the plot that could be farmed by a single family and that could support it economically. And this was indeed the plot that most prospective settlers purchased at government land offices throughout the nineteenth century, before setting out to take possession of it. To this day, when flying over the American West and looking down from forty thousand feet, the vast majority of the regular identical squares one sees stretching from horizon to horizon are a quarter mile by a quarter mile, or exactly 40 acres.[71]

The Spirit and the Grid

None of this, as Jefferson had already concluded, would have been possible as long as the native people of America freely roamed the land. The people for whom every hill and grove was imbued with their stories and beliefs, who considered the land of their ancestors to be a living part of who they were, would never voluntarily acquiesce to the featureless monotony of the gridded land. Nor did the prospects of the Indians settling peacefully into square forty-acre tracts and farming the land alongside their white neighbors seem promising. Jefferson may have believed that the Indians would ultimately accept their full assimilation into white society, but even he conceded that it would take many generations before that happened. For a more realistic assessment of Native American attitudes, he would have done well to pay heed to the words of Tecumseh, the Shawnee war chief who led an intertribal coalition resisting white settlement. "The Great Spirit above knows no boundaries," Tecumseh wrote Jefferson in 1807, "nor will his red people acknowledge any."[72]

This being the case, the gridding and settlement of the land required that the Indians first be cleared out. Jefferson, as we have seen, had acknowledged as much in his private correspondence, where he openly discussed the ways in which this could be accomplished.[73] But troubled, perhaps, by the brutality of such a plan, or perhaps merely by the bad optics it would produce, he never made it official US policy.[74] His successors had fewer qualms. A succession of wars between 1812 and 1818 against Tecumseh's coalition, as well as the Creek and Seminoles, resulted in the mass dispossession of native people. Seeking to regularize the process, President Monroe in 1824 proposed that the Indians be removed and resettled on lands west of Lake Huron, where they would be taught to work the land on small plots, like their white neighbors. An 1823 decision by the Supreme Court that ruled that the Indians did not have title to their lands but only "rights of occupancy" served as the legal foundation for Monroe's proposal.[75]

Monroe's proposal was never approved by Congress, and it was shelved by his successor, John Quincy Adams, who preferred negotiation with the tribes over mass expulsion. But the 1828 election of Andrew Jackson to the presidency breathed new life into the plan. In 1830 Jackson signed into law the "Indian Removal Act," which provided for the systematic uprooting of Indians from their lands in the east and resettlement beyond the western frontier. The law made no mention of forcible removal and in fact referred to "such tribes or nations as may *choose* to exchange the lands where they now reside." But in its actual implementation over the following decades, little "choice" was actually on offer.[76]

Continuing conflict between Indians and white settlers soon made clear how the law would actually be implemented. In the old Northwest, continued encroachment by settlers into tribal lands provoked Indian retaliation, led by the Sauk war chief Black Hawk. The short and bloody war that followed ended with the capture of Black Hawk in August 1832 and the mass expulsion of his followers from their lands. Meanwhile in the Southeast, members of the "Five Civilized Nations"—the Cherokee, Muscogee, Seminole, Chickasaw, and Choctaw—were coming under increasing pressure from their white neighbors who were blatantly and violently infringing on their territory. In response to an appeal by the Cherokee, the Supreme Court ruled that US states, and in particular Georgia, must adhere to the terms of existing treaties that guaranteed the Indians' territorial boundaries, but Jackson openly defied the ruling. He began the systematic expulsion of the Cherokee from Georgia, sending them thousands of miles to the west, and his successors continued the process, applying the same policy and the same methods to the other southeastern tribes. Thousands of exiled Indians died on the long trek westward, which became known as the Trail of Tears.[77]

Removing the Indians from the eastern United States was one thing, but where would they be settled? Jefferson had referred vaguely to lands "beyond the Mississippi" as their destination, but by the 1830s the states of Missouri and Louisiana, and the soon-to-be state of Arkansas, were already established on the west bank of the great river. The Removal Act of 1830 was not much clearer on the matter, and so it was followed in 1834 by the Indian Trade and Intercourse Act, which defined the lands west of Missouri and Arkansas as "Indian Country." A slice of this territory, the act directed, would be allotted to each of the tribes arriving from the east, which "the United States will forever secure and guaranty to them and their heirs and successors."[78]

With the general area of Indian Country now established, the contours of the specific segments allotted to each tribe were also set down. The western boundary of Missouri and Arkansas was set in 1828 as the meridian passing through the mouth of the Kansas River, and this established the eastern border of "Indian Country" as a fixed straight line, running from north to south. Lined up in succession along this line were the Indian nations, some of them local, many transferred from the east: the Omaha, Pawnee, Otoe, Iowa, Kickapoo, Delaware, Kansa, Shawnee, Osage, Cherokee, Creek, and Choctaw. Each tribe was lined along a particular stretch of the line marking the eastern boundary of Indian Country; each was also separated from its neighbors to the north and south by straight boundaries running east to west, intersecting with the eastern boundary at right angles. The tribes that settled in Indian Country, in other words, were confined within strict rectilinear borders aligned with the points of the compass. They had been boxed in, captives of the Great Western Grid.

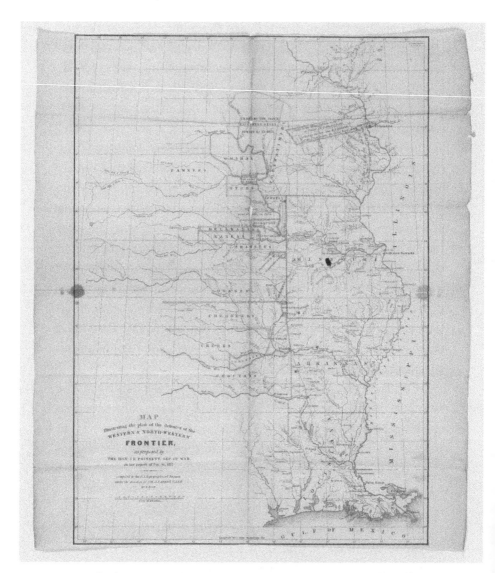

Map of Indian Country showing the rectilinear boundaries of the tribal lands, 1837. From the secretary of war's report to the US Senate on the protection of the western frontier, December 30, 1837. Courtesy of the National Archives.

The tribes had little time to get accustomed to their new circumstances, for Indian Country did not last. In 1830 Congress had promised to "secure and guaranty" the tribal lands "forever," but this commitment, like so many others made to Native Americans, proved to be an empty one. Just as Jefferson had foreseen decades before, relocation to the west served only as a temporary respite for the Indians from the pressures of white encroachment. In time, he had predicted, white settlers would move into the designated Indian territories, new states would be carved out, and the tribes would be pushed even farther westward.[79] In practice, even the respite granted the Indians was partial at best, for they had to contend with a steady trickle of settlers crossing into Indian Country and squatting on the land, correctly assuming that their ownership rights would soon be recognized by the US government. Finally, in 1853, Congress cast away all pretense of honoring its promise, by authorizing the government to acquire land in Indian Country for the express purpose of opening it up to white settlement. A year later the northern two-thirds of the area were carved into the new territories of Kansas and Nebraska, leaving in place only the southern part (the future Oklahoma) as "Indian Country." The work of surveying and settlement within the new territories began in earnest, while the Indians were confined within the ever-tightening bounds of rectangular reservations.[80]

For this was the ultimate triumph of the grid over the Indians. Exiled from their ancestral home where they had acknowledged no borders, they were forced into lands defined to the north, south, and east by straight rectilinear lines. After 1853 their lands were further constricted and shrunk into "reservations," square and rectangular islands of open space surrounded on all sides by regular and precisely measured rectilinear plots. Even then, however, the grid was not done transforming the tribal lands. Year after year it encroached on the reservations, ultimately covering even those last holdouts with its uncompromising rectilinear net of numbered straight lines aligned with the points of the compass. Tecumseh's land of the Great Spirit had been absorbed by Jefferson's mathematical landscape.

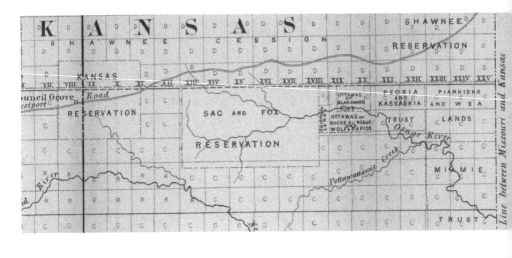

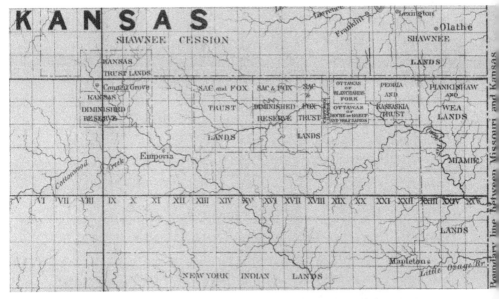

Segments of surveyor maps of Kansas and Nebraska from 1857 (*top*) and 1862
(*bottom*). In 1857, the Kansas reservation and the Sac and Fox reservation were still
open land. By 1862, they had been absorbed by the grid.
Courtesy of the Library of Congress, Geography and Map Division.

Living in Jefferson's Mind

It took a century and a half, but by the 1930s the survey of the lower forty-eight states, or the "contiguous" United States, was complete. To a stunning degree, the western United States had become precisely what Jefferson had dreamed it would be: a vast uniform and undifferentiated space, marked by a vast uniform mathematical grid. It was a continent-sized region with all its riches: mountains and rivers, forests and deserts, giant redwoods and open prairies, roaming grizzlies and herds of buffalo, a region that had been inhabited for millennia by native peoples of rich and diverse cultures. Much of it was ignored in Jefferson's scheme, much of it physically removed, and none of it mattered: it was all recast as an abstract mathematical space, modeled on Newtonian absolute space and marked by the Cartesian coordinates of analytic geometry. The vision in Jefferson's mind in the 1780s had become the reality of the 1930s, and it remains so today.

Why did Jefferson strive to remake American space into a mathematical grid? The reason most often cited by Jefferson's contemporaries is that dividing the land in this manner was a practical and easy way to rapidly settle the vast territory of the American West. Hugh Williamson, for one, who claimed to be the author of the idea, promised Governor Alexander Martin in 1784 that "the plan will prevent innumerable frauds and enable us to save millions."[81] William Grayson, who succeeded Jefferson in chairing the congressional committee on western lands, wrote the following year to assure a skeptical Washington that the plan "is attended with the least possible expence" and that whatever costs were nevertheless incurred would be repaid to Congress "ten fold" through (once again) "preventing fraud."[82] Modern scholars of the grid have generally concurred: according to Hildegard Binder Johnson, the unambiguous tract boundaries made for the "easy" transfer of land from the government to individuals, and Norman Thrower, who compared the long-term development of gridded and ungridded districts in Ohio, noted that there has been far more land litigation in the ungridded regions than in those in which the rectilinear grid prevailed. More recently, Ted Steinberg has argued that the grid is part of the ongoing process of "commodification" that characterizes Americans' relationship to the land.[83]

Nevertheless, practical considerations cannot explain why Jefferson attempted to impose a precise Cartesian grid on the western landscape. They certainly do not explain why generations of Americans would work tirelessly to make this eccentric vision a reality. We have already seen that, as early as 1784, Jefferson's rectilinear plan was launched in the face of strong opposition from Washington and others, who considered it overly rigid, unreasonable, and unnecessary.[84] And indeed, the experience of the nineteenth and twentieth centuries, when western lands were actually divided and distributed according to Jefferson's immaculate Cartesian grid, fully confirmed the warnings of these early critics.[85]

For, as its critics noted from the start, the grid is anything *but* practical. Ignoring hills, valleys, streams, and forests, it imposes its rigid boundaries on an unwilling landscape. As Washington predicted, it creates artificially equal plots out of inherently unequal locations, which differ drastically in land type, quality, and access to water. One settler might, for example, acquire a plot of rich and level farmland, with a stream flowing near the edge, providing a reliable source of water and a suitable site for a mill. A less-fortunate neighbor might end up in possession of a plot of exactly the same shape and size but made up of bone-dry rocky hills. And all for the same price.[86]

Then there are the roads, which in the land of the grid to this day are limited to the boundaries of rectangular plots, following strict north–south and east–west cardinal directions. Since they entirely disregard the geography, they repeatedly go straight up hills and down valleys, turning a drive in even seemingly level terrain into a roller-coaster ride.[87] Perhaps most impractical is the fact that because they ignore the natural contours of the land, the roads do not adjust to creeks and streams that could easily be avoided, or crossed once, with just a minor modification of their route. Instead, they end up crossing the same stream over and over again, requiring numerous bridges

The impractical grid: Roller Coaster Road in northeast Iowa.
From iStock.com/igorkov.

where one, at most, would be necessary. And because of the odd angles at which the straight-arrow roads cross meandering creeks, those bridges can be quite lengthy—and expensive.[88]

If this were a price to pay for the rapid transfer of land to settlers, it could, perhaps, be justified. This might have been the case if prospective settlers had purchased their square plots on a map in a distant office in a major eastern metropolitan area before setting out on the long trek west to take possession of it. But this was not the case. The settlers did not purchase their plots in advance, sight unseen, but first traveled to the region where they hoped to buy land. They then bought their plot from a nearby land office, after having a chance to view their future holdings.[89]

At no step in this process did the perfect Jeffersonian grid make the transfer of land from the government to the settlers simpler, easier, or faster. The local surveyors could have divided the land sensibly and equitably if they had been freed from the demands of the tyrannical grid and could have taken into account local conditions. They would likely have done it faster as well, since the need to establish rectilinear lines so precisely that they would join with all others across the continent presented an extreme challenge and served no purpose locally.[90] Finally, since the settlers bought the land locally, and not from afar, they could have done the same regardless of the shape of the lot—rectilinear or otherwise—especially since they had the opportunity to actually inspect the land and decide whether it answered to their requirements.

All of which is to say that the grid was, and largely remains, eminently impractical. The lands of the West could have been surveyed and settled at least as fast and at least as easily without carving the land into a precise checkerboard pattern. The results for the settlers, furthermore, would have been equally good—if not better. No one, after all, had considered creating a grid when the original colonies on the Atlantic Seaboard were settled, and no other European colonial project, in America or elsewhere, had ever attempted one. From the French in New France and Louisiana to the Spanish in the Americas and the British in Australia, no one thought it useful, efficient, or practical to carve the land into a Cartesian grid. No one, that is, except Thomas Jefferson.

Why did Jefferson do this? Why did he insist on imposing a pure mathematical grid on a recalcitrant continent? Why did American administrations from Jefferson onward embark on such a titanic undertaking? Why did they create, from scratch, a massive administrative apparatus and maintain it from generation to generation over a century and a half—an undertaking, furthermore, that in its early decades was conducted at the very limits of technological feasibility? It was immensely hard, immensely expensive, entirely unnecessary, and fully impractical. So why did they do it?

The answer is that the great grid is not, and never was, a practical construct. It was, to the contrary, an ideological one. The massive, unified Cartesian grid, made up of precise mathematical lines running north–south and east–west over the western United States, is there not because it is useful but because it encodes a particular vision of the United States. A vision that Jefferson and his successors inscribed on the land. A vision that, to a large extent, the grid still carries to this day. In Jefferson's vision, America was a blank slate on which new men, acting freely, will forge a new society and a new nation. And, as Jefferson had learned from Newton and Locke, no space was as blank, as empty, and as open to action and possibility as simple, empty, and unresisting mathematical space. For Jefferson, America was already that empty land of freedom and opportunity; the great grid merely marked it as such and when necessary made it so, by removing unwelcome reminders of its natural and human history.

In part, the grid was a means for making the United States into the nation that Jefferson insisted it should be. It transformed the complex landscape of the American West into simple agricultural land, without history, tradition, or (to the extent possible) inconvenient natural features. In doing so it opened the land for settlement by those who sought to escape the constraints and hierarchies of Europe and the Atlantic states and build their lives as free people in the open and unconstrained West. These were the very people whom Jefferson thought "the chosen people of God," free tillers of the soil, who were the most virtuous of men. Because the grid made the land into a uniform empty space, these men will be free of all constraint. Because of the grid's standard plots, these men will be equal. Because of the spacing of plots, which placed each family at a distance from all others, these men will be independent. And because of the rigors of surviving and making a living alone, on untrammeled virgin land, these men would be hardworking and enterprising. Free, equal, independent, and enterprising—those, Jefferson believed, were the new men who would make the new nation. The perfect mathematical grid will turn them from dream to reality.[91]

But the grid was not just a tool for making new Americans; it was also, in and of itself, an embodiment of Jefferson's American ideal. Consider, for comparison, how the rigid Euclidean geometries of the gardens of Versailles embodied the ideals of the French monarchy.

The strict, irresistible, and unchallengeable hierarchies of Louis XIV's state found perfect expression in the strict geometrical patterns of his royal gardens, all converging to the royal palace at the top of the valley. At Versailles, every stone, tree, and blade of grass, as well as every person, had its unique God-given place in a single great hierarchy culminating in the person of the king. To one walking the paths of Versailles, royal absolutism becomes not the boast of an arrogant prince but an inescapable part of the universal order, as true and unchallengeable as a geometrical proof.[92]

And just as the Euclidean geometries of Versailles stood for a fixed and unyielding hierarchy, so did the Cartesian geometries of the American West point to a very different kind of social and political order: egalitarian, dynamic, democratic, and filled with the promise of yet-to-be-fulfilled potential. Without boundaries or a center, the grid is homogenous—all points within it are the same, none superior or inferior to any other. If at Versailles all eyes are drawn to the king's seat, from whence all power emanates, in the grid nothing captures the eye, all squares are created equal, and power—if it is present at all—is uniformly diffused throughout space.

Versailles, with its immovable triangles, is a vision of a single inescapable social and political order. The great grid, in contrast, defines no particular order at all: like graph paper before it is inscribed with a graph or curve, it defines an empty space of possibility, without determining which truths will actually be inscribed in its unbounded and unresisting expanse. Consequently, whereas Versailles puts everything and everyone in their appointed place, the grid is nothing but a call to action, an invitation to every inhabitant to make their mark on this vast empty plane. To American settlers the interminable grid, with its endless boring monotony, was a place of endless possibility. It was where freedom reigned, where no man lorded over his neighbor, and where they could create their own little private kingdom in the vast void. It was where all these free men, each in his own domain, would join to forge the Empire of Liberty.

This was Jefferson's vision for the new land, and it was bold to the point of rashness. It was, in most ways, the exact opposite of what one might expect of a nation coming into possession of a vast unexplored territory. Normally settlers and colonizers would seek to learn as much as they could about the new land. They would study its climate, the different types of soil, the natural flora and fauna, the mountains and rivers. And they would, inevitably, familiarize themselves as much as possible with the human inhabitants as well. When moving into the territory, they would do their best to accommodate the natural and human conditions and seek out the best soil and climate and the most suitable environment for settlement. And that is indeed how the settlement of the West proceeded before the implementation of the continental grid: settlers moved into new territories they thought most promising and then marked the irregular boundaries of their claims. Even George Washington, who owned large tracts of western land, could not imagine things any other way.[93]

Jefferson knew this as well. His *Notes on the State of Virginia* is chockfull of information on the unique topography, flora, and fauna of the state, as well as its human inhabitants. His instructions to Lewis and Clark as they set out on their expedition to the West directed them to collect detailed information on the land and its people.[94] But what Jefferson first proposed, and the national survey ultimately implemented, was the exact opposite. Not only did the uniform grid ignore the natural landscape; in as much as possible, it sought to erase it. Mountains, rivers, creeks, and forests were not realities to be used or accommodated but rather inconveniences to be ignored and, when necessary, overcome. Indeed, if Jefferson had had his way, neither river nor mountain would have modified the perfect squares with which he sought to cover the continent. His goal was to transform the land into an immaculate mathematical grid while erasing the imprint of natural, human, or historical conditions. To an astounding degree, he succeeded. The western two-thirds of the United States exist today much as they appeared in Jefferson's mind: a formal homogenous landscape of straight lines and right angles, modeled on the mathematical space of analytic geometry.

That Jefferson conceived of this plan speaks to his idealism and his habit of thinking in broad abstractions rather than getting bogged down in practical details. These were the habits of mind of an Enlightenment philosopher and a mathematician, and they gave us the Declaration of Independence, with its bold assertion of human equality and universal rights. But the fact that he managed to implement his plan to the degree that he did also speaks to something else: in it we hear the echoes of the optimism, if not arrogance, of a young nation coming into being. Having defeated their British overlords and gained their independence, Americans had acquired a sense of their own omnipotence in the face of not only human enemies but also natural constraints. "Brothers, you seem to grow proud because you have overthrown the king of England," a perceptive Shawnee spokesman declared to the Americans at an abortive peace conference in 1785.[95] And so it was. With burgeoning self-confidence, the free citizens of the young republic set out to shape the world as they saw fit. Neither natural nor human obstacle would stand in their way as they transformed a rich and diverse continent into a formal mathematical landscape of precise and identical squares.

To Jefferson and those who shared his vision of America, the grid was key to molding a new people and a new nation. It would make men who were free and equal, independent and unconstrained by nature or human society, and call on them to build a new nation by the sweat of their brow. Such men, he believed, who tilled their own fields and raised their own crops, would love freedom, owe nothing to those who considered themselves their betters, and preserve the republic against its enemies for all time. Such was Jefferson's vision and dream for the United States, and at its heart was the great monotonous grid that would cover the continent.

That Jefferson succeeded far beyond what any objective observer had a right to expect when he first floated his proposal, there can be no doubt. The American West of today, with its straight lines, right angles, and innumerable identical squares is the physical incarnation of Jefferson's idea, as he presented it to Congress in 1784. It is not a stretch to say that those like me, who live in the western United States, are living inside Jefferson's mind.

But did he succeed in his broader goal, of shaping not just the land but the people and the country? This question is more difficult to answer because it is undeniable that today's United States is vastly different from the one Jefferson envisioned more than two centuries ago. Jefferson's utopian democracy was to be made up first and foremost of virtuous yeomen farmers, the very men who would settle the West and grow their crops on their square plot of land. For the great metropolises of the Union, the seats of commerce, finance, and manufacture, Jefferson had nothing but contempt. And yet in the modern United States the majority of Americans live in cities, depraved or not, and their occupations have little to do with daily toil in the fields. For better or worse, today's United States is not Jefferson's idealized vision of a federation of virtuous independent farmers.

So did Jefferson fail? Did the great grid he initiated, whose implementation took a century and a half, result in nothing more than turning a rich and varied natural landscape into a boring monotonous agricultural one? Speaking as one who came to the United States from a different place and culture, I believe that is not the case either. For it seems to me that the human virtues that Jefferson associated with the rectilinear grid have indeed become core parts of the way Americans view themselves and their place in the world. Alexis de Tocqueville described Americans as having "a restless spirit, immoderate desire for wealth, and an extreme love of independence."[96] They are also, I would add, fiercely egalitarian, suspicious of elitist pretensions—whether based on birth, wealth, or education—individualistic, and free from the claims of family and clan to an extent hardly imaginable in most other countries. Whether on the right or the left, Americans insist on their freedoms and guard them vigilantly, forever casting a wary eye on those with political or economic power. And even in an age of deepening pessimism, Americans still believe that their country should be a land of limitless opportunity, where with hard work and an enterprising spirit even the poorest and most disadvantaged can succeed. It is a profoundly American story, and it resonates deeply. The current cultural malaise, which has Americans openly denouncing both their past and their present, does not stem from a rejection of this narrative. It comes, rather, from a sense, shared by both the right and the left, that the country is failing to live up to it.

Speaking broadly of American "attitudes" and "ideals" is inevitably a hazardous venture.[97] There are surely millions who identify as Americans but in no way fit such general descriptions. And yet, as someone who grew up outside the United States and has made his home in California, it is my belief that these characterizations truly do capture something fundamental about the land and its people. Surely not all Americans are individualistic, egalitarian, hardworking, and enterprising. But for all their wildly differing backgrounds, it's fair to say that the vast majority believe that those very traits are fundamental to what makes one an "American." Jefferson hoped to instill these values in his people and found a nation like no other. In true American fashion, he succeeded beyond what he could possibly have had a right to expect.

Despite being an avid letter writer and the author of the Declaration of Independence, Jefferson published little during his lifetime. If he is nevertheless one of the most influential figures in American history, it is not solely because of his writings, his political activity, or even his presidency. It is also because Jefferson wrote down his legacy not just on paper but on the vast expanse of the American continent. The Great American Grid is Jefferson's legacy written on great forests and open prairies, and it is a legacy that has shaped Americans' ideals and self-understanding. The vision of America as the land of freedom and opportunity is inscribed in the limitless open spaces of Jefferson's gridded West. And whether they live in the city or on the prairie, it is a land that Americans, in their minds and in their hearts, still inhabit today.

Rectilinear Cities

The Island of Hills

"We know nothing more frightful than to look at a plan of New York, and fancy what it will be when the whole island is built upon," the *Athenaeum*, the popular London journal of literature, science, and the arts, declared in the summer of 1849. "Never was so noble a gift of nature so sacrificed to mathematical precision and utilitarian caprice," complained the anonymous author. "Let the reader imagine a dozen of Harley-streets, Baker-streets, and Edgeware-roads, all parallel to each other, and extending in a straight line from six to ten miles in length, without a Hyde Park, a Regent's Park, or a Hampstead intervening, and intersected at unvarying intervals by a couple of hundred cross streets, all the ditto of each other,—and he will have an idea of what New York is to be."[1]

For all its hyperbole, the *Athenaeum* was not far wrong, for by 1849 New York City was closing in on its fourth decade of implementing an unprecedented program of urban expansion. In 1811 the New York State legislature approved a municipal plan for the city that envisioned carving up the island of Manhattan into a regular Cartesian grid composed of numbered streets and avenues intersecting at right angles. Ever since, the municipal authorities have been busy turning the plan from a paper diagram into reality on the ground. The hills and valleys where, only a few decades before, Washington's Continental Army had faced the British redcoats were leveled, the woods chopped down, and the creeks dammed. Broad and straight north–south avenues numbered from 1 to 11 were surveyed and marked from one end of the island to the other. Narrower, more numerous east–west streets, stretching from the East River to the Hudson, were regularly laid out, beginning with 14th Street, which marked the southern boundary of the grid, and moving northward. Year by year and decade by decade, a flat two-dimensional grid was settling like an all-encompassing net over the rugged island at the mouth of the Hudson River, which the Lenape people had named Manahatta, the "Island of Hills."[2]

As the grid expanded, the city followed, moving northward from its origins at the southern tip of Manhattan. By midcentury, when the *Athenaeum* issued its grim warning, all the blocks from 14th to 38th Street were fully built up and populated. Certain neighborhoods on the east and west sides stretched even farther north, to around 51st Street, where they were separated by open spaces in which streets and avenues were marked out but had yet to be built. Meanwhile, New York's population was growing by leaps and bounds, powering the rapid expansion northward: a city of 32,000 in 1790 had tripled to 96,000 by 1810, tripled again to 312,000 by 1840, and reached more than half a million by midcentury, making it far and away the largest metropolis in the United States. Whether one admired the rationality and functionality of the plan or despised it as did the anonymous writer for the *Athenaeum*, the future of New York City was clear to see: it would not be long before the city's population exceeded one million, and the grid would cover the entire island of Manhattan.[3]

The systematic gridding of Manhattan throughout the nineteenth century cannot but bring to mind a very similar process taking place on the far reaches of the continent at the very same time. Even as municipal surveyors, under the direction of John Randel Jr., were spreading across the island of Manhattan intent on realizing the Commissioners' Plan for New York, an army of federal surveyors under the direction of Jared Mansfield was fanning across the western

United States to realize Jefferson's plan for the American West. There were, to be sure, significant differences: In New York, the surveyors were marking avenues and streets; in the West, they were measuring ranges and townships. And whereas the basic unit in the western survey was the six-mile square, Randel's surveyors divided the island into rectangular blocks 260 feet in breadth and around 800 feet in length, depending on their location. On the great western plains, the survey lines were abstract markers on the land, straight lines marked on a map that also served as boundaries between properties and fields. In New York, the lines became actual streets, 60 to 100 feet wide and ultimately paved with cobblestones or asphalt, occupying land that had been physically leveled to create a flat surface. And then there was the difference in scale: whereas the continental survey, when completed, divided up close to 1.5 billion acres of land, the entire land area of Manhattan is just the hundred thousandth part of that—less than 15,000 acres.

Yet for all these differences, there is no denying that the guiding vision behind the two surveys was the same. In both cases, a rich and varied landscape was reconfigured into a uniform two-dimensional space marked only by an abstract mathematical grid. And just as with Cartesian coordinates, in both cases each and every location on this homogenous empty landscape was uniquely determined as the intersection of two lines meeting at a right angle. In the great western survey, each six-mile square is defined by an east–west "range" and a north–south "township," each of them numbered according to its distance from the intersection of a principal meridian and its base line. And in Manhattan, to this day, each point is defined as the intersection of a north–south avenue with an east–west street, both of them numbered in succession. In effect, both the great continental survey and the Manhattan street survey transformed their landscapes into Cartesian grids, modeled on the mathematical space of analytic geometry.[4]

Two different grids were systematically charted onto the landscape from year to year and decade to decade throughout the nineteenth century. One was rural; the other urban. One covered a continent; the other a small coastal island. Yet despite their differences in character and scale, they shared a single vision. Both sought to transform a complex terrain steeped in human and natural history into a mathematical blank slate with no past, no history, no human limitations, and no natural constraints. Both Manhattan and the American West, each in its own way, would become a landscape of limitless possibilities, passive, open, and accepting of whatever was imposed upon it. It was a land that enterprising Americans, unconstrained by human traditions or natural conditions, will shape as they saw fit, where they will freely pursue their dreams while building a nation like no other. And so as the great Jeffersonian survey moved systematically west, incorporating vast territories into its relentlessly monotonous scheme, the Manhattan survey moved systematically north. It was like a small and busy microcosm of the great grid taking shape beyond the Appalachians.

There is irony here, for Jefferson, the prophet and mover of the Great Western Grid, despised cities. Whereas the life of a farmer enhanced virtue, independence, and freedom, city life led only to corruption, moral degeneracy, and dependence. "A city life offers you indeed more means of dissipating time," he wrote a friend toward the end of his life, "but more frequent, also, and more painful objects of vice and wretchedness." And whereas all cities were deeply suspect, none, in Jefferson's esteem, had sunk lower than the most populous among them: "New York," he wrote disgustedly, "seems to be a Cloacina [cesspool] of all the depravities of human nature."[5] If independent yeomen farming on the open range were, according to Jefferson, honest, virtuous, and freedom loving, then city dwellers were the opposite—corrupt, selfish, and subservient to power. "The mobs of the great cities add just so much to the support of pure government, as sores do to the strength of the human body," he had written as far back as the 1780s, and he never wavered from this belief.[6] Cities, in the Virginian's opinion, corrupted the soul, and the greater the city, the greater its corruption. True republicans must steer clear of metropolises like New York as much as possible or risk grave danger to their republican spirit.

Versailles on the Potomac

Yet cities had to be built—even Jefferson conceded as much. The challenge, as he saw it, was to design them in ways that would limit their malign influence. And so in 1791, while serving as secretary of state in Washington's administration, he jumped at the opportunity to put his stamp on the new federal capital, the city that would ultimately become Washington, DC. As envisioned at the time, it would be the largest and grandest of all American cities, far eclipsing commercial hubs such as Philadelphia and New York. If Jefferson could mold it to his vision, it would serve as a model to future cities that were sure to rise farther west, and hopefully contain their damage to the body politic.[7]

A federal capital had been decreed in the Constitution, but as no land had been set aside for it, the city's location became the focus of intense bickering between the states. It was only when Jefferson, in the summer of 1790, brokered a compromise between the northern and southern factions that Congress finally approved the site on the north bank of the Potomac River. Given his role in securing the capital's location, Jefferson likely expected to have a major say in planning the new city, but President Washington had other plans. As chief designer, he selected his comrade in arms Pierre Charles L'Enfant (1754–1825), who had come over from France in 1777 to fight in the cause of liberty.[8]

L'Enfant was not only a Frenchman but a courtier, the son of a painter to King Louis XV. He had learned his craft at the great royal gardens for which France was famous, and his design for the American capital bears their unmistakable imprint. At Versailles, the grandest and most famous of them all, arrow-straight "allées" intersect at precise angles to forge an elaborate geometrical landscape of triangles, rhombuses, and rectangles. Together they join into broad boulevards that converge symmetrically up the valley onto the king's palace, the center of all power and authority. The message to anyone wandering the paths of the gardens of Versailles is clear: the supremacy of the kings of France is as unchallengeable and irrefutable as the truths of geometry. Like geometry itself, it is a manifestation of the deep order of the universe.

And so, when he was charged with designing the federal capital, L'Enfant's path was clear: he would apply the principles that had made Versailles the most famous royal seat in the world to the city on the banks of the Potomac. He promised that, when built, Washington would be a capital worthy of the young American "empire" and would outshine all the great capitals of Europe.[9] Yet transferring the vision of Versailles from France to America proved to be a challenge. Versailles, after all, was a royal estate, built to embody the absolutist ideal that the king was the one and only fount of all power and legitimacy. The city of Washington (as it would soon be named), in contrast, was the seat of a republican government, founded on the newly ratified Constitution. Rather than embodying a single source of authority, the government of the United States consisted of multiple sources of power, all carefully balanced against one another: the president against Congress, the House against the Senate, the federal government against the states. Could the formal patterns of a French royal palace be used in an American republican capital? The strict geometries of Versailles, after all, were so closely associated with absolute monarchy that royal courts across Europe had rushed to create their own version of Louis XIV's famous garden.[10] Implementing them in the capital of the United States, it seemed, would undermine the young republic's foundational principles.

L'Enfant, however, was undeterred. In his plan for the capital, he included all the familiar trappings of Versailles: arrow-straight arteries crisscrossing the city and intersecting at precise angles; star-shaped plazas radiating broad avenues in all directions; a "grand boulevard" (familiar today as the National Mall) rising up toward the palace on the hill (also known as the Capitol). Like Versailles, the federal capital was to be a tapestry of geometrical shapes fitted seamlessly together. And yet the overall effect of L'Enfant's design was anything but a celebration of monarchical absolutism.

In place of a single fount of unrivaled authority, L'Enfant's city would have two: the "Federal House," known to us as the Capitol, and the "President's Palace," or, to us, the White House. Each of these is, in effect, its own great palace of Versailles. Each stands grandly at the top of sloping gardens, each is the focal point of grand boulevards converging from all directions, and each entirely dominates its surroundings. And yet the two poles of power are not disconnected but carefully balanced, pushing and pulling against each other at the two ends of a broad boulevard, which L'Enfant named Pennsylvania Avenue. The constitutional balance of power between the executive and legislative branches of the government is written into the federal capital in the language of geometry.

And that is not all. For beyond the great centers of federal power, and reaching into the outskirts of the capital, the city is overlain by a network of fifteen smaller local centers. Each of these is a star-shaped plaza, with linear streets radiating in all directions, and each dominates its local surroundings just as the Federal House and the President's Palace dominate the city as a whole. L'Enfant was not coy about the meaning of these local centers: every plaza, in his design, was to be named after one of the fifteen states that formed the Union at the time and decorated with monuments unique to it. In L'Enfant's capital, as in the Union itself, each state dominates locally but cannot challenge the federal institutions that tower over all of them. Yet, in both the city and in the Constitution, the power of the states taken together constrains the power of the federal government, like an overhanging net that they can never escape. At Versailles, geometrical landscaping made the rule of the Bourbon kings appear as irrevocable and unchallengeable as the truths of geometry; on the banks of the Potomac, L'Enfant did the same for the rule of Congress, the president, and the states. On the streets and plazas of the federal capital, the constitutional division of powers became a manifestation of universal geometrical order.

Like Versailles, L'Enfant's city was a political ideal written in stone, pavement, and greenery. But whereas in monarchical Versailles this ideal was at its core a simple one, in the city of Washington it was remarkably complex, balancing competing poles of power and the center against the periphery. That L'Enfant accomplished this with

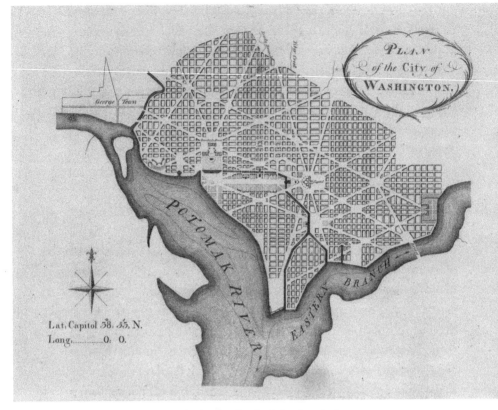

L'Enfant's plan for the federal capital, 1791.
Courtesy of the Library of Congress, Geography and Map Division.

such elegance and grace, designing a beautiful yet functional city that was also a philosophical and political statement, seems almost miraculous. It is surely one of the greatest triumphs of urban design anytime, anywhere. And Thomas Jefferson hated it.

The Frenchman and the Secretary

We know this because Thomas Jefferson had proposed his own plans for the federal capital, and they could hardly have been more different. As early as November 1790, Jefferson had enclosed a sketch of the future federal capital in a report to President Washington about the progress

made in securing a site for the new city. At the time, difficulties in ne-
gotiating the purchase of the land near Georgetown had convinced Jefferson that the capital would likely be built farther to the south and
east, at the point where the Anacostia (or "Eastern Branch") flows into
the Potomac River. But a few months later there was a breakthrough in
negotiations with landowners, and the capital site was moved farther to
the west. Jefferson quickly sketched a plan suitable for the new location
and delivered it to Washington at the end of March 1791.

By American standards, Jefferson's plan for the federal city was
unquestionably grand. Along with his first sketch Jefferson, proposed
that "no street be narrower than 100. feet, with foot-ways of 15. Where
a street is long & level, it might be 120. feet wide." As for the size of
the capital, he estimated that "1500 Acres would be required in the
whole, to wit, about 300 acres for public buildings, walks, &c, and
1200 Acres to be divided into quarter acre lots."[11] Jefferson did not offer precise dimensions for his second sketch, but the city is depicted
as even larger than in the earlier proposal, covering roughly 2,000
acres. This at a time when Philadelphia, the largest city in the Union,
covered less than 1,000 acres and no city in the land sported streets
and avenues on the scale that Jefferson proposed.[12]

Jefferson's two plans differ in location and size, but even a single
glance at them reveals that their core design is the same: Whether on
the banks of the Anacostia or the Potomac, Jefferson was proposing
a city defined by a strict and regular rectilinear grid. Long avenues
would stretch lengthwise, from one end of the city to the other, while
shorter streets, also traversing the entire city between opposite sides,
would intersect them at right angles and regular intervals. The end
result would be a city designed as a uniform plane, made up entirely
of square or rectangular blocks, all precisely the same size and shape
and all lined up next to one another. If L'Enfant's city is a dazzling
jigsaw puzzle of geometrical figures, dramatically arranged to form
a city landscape, Jefferson's city is a uniform checkerboard in which
all blocks are not just equal but identical.

When Washington forwarded Jefferson's proposals to L'Enfant, the
Frenchman erupted in indignation. "A plan of this sort," he declared

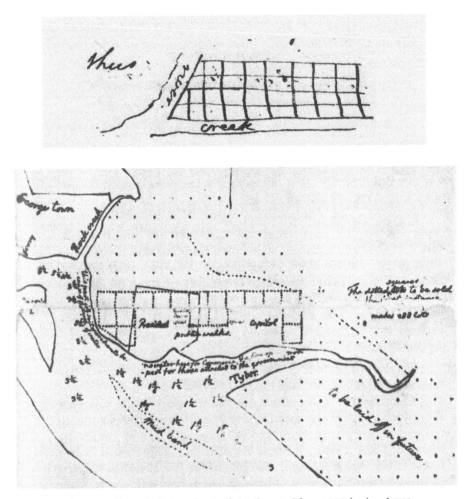

Thomas Jefferson's designs for the federal capital from 1790 (*top*) and 1791
(*bottom*). Courtesy of the Library of Congress, Manuscript Division.

to Washington, "must be defective, and it never would answer for any of the spots proposed for the Federal City . . . it would absolutely annihilate every of the advantages enumerated and . . . alone injure the success of the undertaking." Both "tiresome and insipid," he continued, it was the brainchild of "some cool imagination wanting a sense of the really grand and truly beautiful."[13]

What was it about Jefferson's proposal that so enraged L'Enfant? So much so that it led him to insult the second most powerful man in the land? It might, perhaps, have been the size of Jefferson's proposed city, which, though grand by American standards, still covered only one-third of the area of the Frenchman's proposal. But if it had been merely a question of size, L'Enfant could have simply suggested expanding the city. Indeed, it is one of the advantages of a gridded street plan that it can be easily expanded to cover any proposed area. Was it, perchance, that he considered the scale of the federal buildings in Jefferson's city to be insufficiently grand? Indeed, L'Enfant's own suggestions for the Federal House and the President's Palace were on the scale of Versailles and the Tuileries, well in excess of what the secretary was proposing. Yet the seats of government in Jefferson's plan were, in truth, grand enough, their grounds comprising up to one-fifth of the entire city. If L'Enfant sought to expand them, to match the larger area of his city, he could have suggested as much.

The true reason why L'Enfant was so outraged by Jefferson's plan was not a matter of size or grandeur but of something even more basic, something so fundamental as to be almost invisible: the plan's geometry. L'Enfant's city, like Versailles, is a rigorously Euclidean landscape, in which everything is exactly as it must be. Try moving one of the great avenues in L'Enfant's design a little to the north or south or altering its angle ever so slightly, and it becomes immediately apparent that it is practically impossible. Each boulevard is held in place by an elaborate network of interlocking plazas and intersections. Any attempt to move a single boulevard will disrupt the entire interconnected network, altering angles at intersections throughout the city and effectively destroying the elegant star-shaped plazas. L'Enfant's plan, in other words, is unchangeable; it is a world in which everything is inflexible and unalterable, a world whose intrinsic order is as unchallengeable and irrefutable as Euclidean geometry itself.

In Jefferson's city there was nothing of the sort. It wasn't just that the layout of Jefferson's city spoke of a more egalitarian order than the Frenchman would have preferred, although it surely did. It was, rather, that Jefferson's plan for the capital promoted no particular order at all. The gridded city has no focal points, no star-shaped plazas, and no grand avenues or parks extending from them. It has no center and no hierarchy, just a uniform rectilinear landscape from end to end. It is, in essence, the geometry of an empty mathematical landscape in which anything can be created—or nothing at all. L'Enfant's design was the Constitution set in stone, a political order made real, physical, and as irrevocable as geometrical truth. Jefferson's design was empty Newtonian space, a tabula rasa on which anything could be inscribed, erased, and reinscribed again in a completely different pattern.

L'Enfant would have been wise to contain his fury at Jefferson and his plan. For the Virginian, he would soon discover, was a wily adversary. Knowing that L'Enfant, for the moment, had Washington's backing, Jefferson bided his time. For the next several months, he engaged in polite correspondence with the Frenchman while quietly working to undermine his position. When the increasingly isolated L'Enfant imprudently ordered the demolition of a house that stood in the way of his designs, Jefferson pounced: "It having been found impracticable to employ Major L'Enfant in that degree of subordination which was lawful and proper," he wrote icily to the city commissioners in March 1792, "he has been notified that his services were at an end."[14] And that was that. Though he lived for another thirty-three years, the brilliant Frenchman was never entrusted with a public commission again.[15]

With the prickly Frenchman safely out of the way, Jefferson likely hoped that his own plans for the capital would be given renewed consideration. But it was not to be. Ever anxious that the Union might devolve into a melee of squabbling states, Washington was committed to building a capital that would be the pride of all Americans. A brilliant capital, he believed, was neither a superfluous indulgence for a young and debt-ridden nation nor a presidential vanity project for the city that would bear his name. It was, rather, a pillar of the republic, essential for its long-term survival. And so, even after L'Enfant's disgrace and effective exile, the president remained unshaken in his determination to implement his sparkling design. And when Washington had made up his mind, there was little even Jefferson could do about it.

Not that he didn't try. Indeed, over the following years, Jefferson did his best to obscure and limit the capital's grandeur. He complained that the scale of L'Enfant's federal buildings and monuments was unsuitable for a republic and worked behind the scenes to reduce the size and magnificence of the houses of Congress and the presidential palace. When Washington died, Jefferson made sure that the plan of burying him in a grand mausoleum on Capitol Hill never came to fruition.[16] And during his own presidency, when he lived in Washington for eight years, he cut back drastically on construction funds for the city, leaving it a semirural landscape of woods and country roads, interrupted by the occasional national monument. But in the end it was all for naught. Year after year and decade after decade, the contours of L'Enfant's design slowly emerged from the swampy banks of the Potomac, creating the city we know today. It is, as anyone who has visited Washington, DC, knows, an elegant geometrical city of grand monuments, star-shaped plazas, and broad avenues converging on the centers of federal power, just as its designer envisioned it. It is L'Enfant's city, not Jefferson's.

Yet even though Jefferson lost the battle for the federal capital, in the war over the proper design of American cities he was still the runaway winner. For even a cursory look at American cities reveals that Washington, DC, a dazzling capital of grand boulevards converging on grand monuments and towering edifices, is very much the exception. The basic street plan of the vast majority of American cities is not an elaborate and carefully constructed geometrical pattern à la L'Enfant but a simple, strict, and regular rectilinear grid. This, of course, is as Jefferson would have wished, for the best street plan for a city, in his eyes, was the one closest to the pattern he was imposing on the American West. And so, despite his scathing condemnation of urban life, Jefferson is not only the originator of the Great Western Grid. He is also, to a considerable extent, the father of American city planning.

Jefferson never wavered in his views on urban design. Fifteen years after proposing a grid plan for the federal capital, he was still advocating for what he called the "chequer board" plan, in which built-up squares would alternate with open space. He promised to apply it to the city of New Orleans, recently acquired from France in the Louisiana Purchase. "The atmosphere of such a town would be like that of the country," he wrote, invoking the highest compliment he believed any town could aspire to.[17] And when in 1803 William Henry Harrison, governor of the Indiana Territory, wrote to inform him of the founding of a new and important town to be named "Jeffersonville," the president seemed to consider it a dubious honor. His main response was to encourage the governor to follow his suggestion to lay out the town in a checkerboard plan. In Jefferson's view, towns and cities should be avoided as much as possible. If they nevertheless proved a necessary evil, then they should be made to look as much as possible like his idealized countryside: sparsely settled and following a strict grid pattern.[18]

The Bastard Grid

Jefferson's hostility to urban life did not prevent cities and towns from quickly sprouting up in every territory settled by Americans and claimed by the United States. As settlers pushed the frontier ever westward throughout the nineteenth century, they left in their wake not only farms and ranches but also hundreds of small towns, numerous cities, and a handful of large metropolises. Jefferson could not have been pleased with this blight on the pristine land he had intended for virtuous tillers of the soil. But he could, perhaps, take consolation in the fact that almost invariably the cities took on his preferred street plan. From Cleveland to Chicago to Denver, and on to San Francisco and Los Angeles, the cities of the West, with few exceptions, were built as uniform urban grids. In some instances, as in Jeffersonville, this may have reflected deference to Jefferson and his views on urban planning. In other instances, the regular grid plan served the interests of land speculators, who wanted to measure and sell identical urban lots of standard size. But the fundamental reason that western cities were built according to the grid plan was simple: because of the way the rectilinear continental survey was set up, they had little choice.

The western grid, as conceived by Jefferson and later revised by Congress, made allowances for certain civic institutions. Within each six-mile square township, a single one-mile square section was set aside for "religion," another section was reserved for public schools, and four additional sections were to be kept in government hands for the time being, to be used as circumstances required.[19] What Jefferson's plan did not provide for was any commercial, manufacturing, or residential areas—or in other words, towns. It is a strange omission. Considering the scope and ambition of the grid, it is hard to imagine that Jefferson and his associates in Congress did not foresee the need for urban centers, even modest ones such as local market towns. Such towns, after all, had always been part of even the most rural landscapes in America and elsewhere. Yet such was Jefferson's antipathy to any form of urban life that he made absolutely no allowance for it anywhere in his idealized rectilinear landscape. The American West in its entirety, as far as he was concerned, was to be divided into uniform squares that would be cultivated by free yeomen farmers. If some towns or cities nevertheless emerged there, they were an unwelcome intrusion on virtuous rural life, and Jefferson would do nothing to help them along.[20]

It may have been an oversight born of profound distaste, or it may have been deliberate sabotage on Jefferson's part, but either way the consequences for western towns were decisive. Lacking a place to call their own, in which an urban plan could be envisioned, laid out, and implemented, the towns simply accommodated themselves to the realities on the ground: and that, throughout the American West, meant conforming to the universal ever-present grid. Where towns grew up, their main streets were the north–south and east–west lines of the grid, which already served as both boundaries between plots and country roads. With the main traffic arteries predetermined, the orientation to the points of the compass unalterable, and the surrounding landscape already carved up in a uniform grid, the rest of the town fell into place. Almost inevitably, the minor streets conformed themselves to the directions and distances of the major ones, creating the familiar pattern of the gridded American town. Surprisingly, the standard familiar layout of American towns is not the result of careful planning but of its opposite—deliberate neglect.

Later Congressional legislation did try to correct Jefferson's "oversight" by setting aside half-section plots of half a mile by a mile for towns. As a result, some of the larger settlements, including towns like Columbus, Cincinnati, Chicago, and Omaha that soon grew into cities, did not simply emerge at the intersection of rectilinear country roads but were laid out in advance according to a plan. By this time, however, the standard pattern of town layouts had been set throughout the western United States. Even as they grew into great cities and metropolises, the urban centers of the West forever bore the imprint of what they had been at their birth—the bastard offspring of the great Jeffersonian grid.[21]

The Ancient Scheme

Urban grids, to be sure, were not born in nineteenth-century America. The basic scheme of parallel streets intersected at right angles by parallel streets has a long and storied history and can be traced in some form to the beginnings of urban life. The earliest known example of the pattern is found in the ancient cities of Mohenjo-Daro and Harappa in the Indus Valley of northern India, founded in the third century BCE and mysteriously abandoned a millennium later. Little is known of the Indus Valley Civilization to which they belonged, and no trace is left of their history or original names. But the size and careful design of the cities suggest the presence of a strong central authority and indicate that they served as administrative centers for a large surrounding region. Some of the streets in each city are straight and oriented to the points of the compass—north–south and east–west—creating a pattern of blocks of roughly equal size. Many other streets, however, seem to have grown organically and do not conform to an overall pattern, grid-like or otherwise.[22] We do not know whether Mohenjo-Daro and Harappa were part of a broader tradition of urban planning or whether other cities in the region followed their example. For the next known use of the urban grid took place not on the Indian subcontinent but in Mesopotamia, where the great city of Babylon was laid out in a rough grid pattern, possibly as early as the reign of Hammurabi in the eighteenth century BCE. At least one Assyrian king, Sargon II (722–705), followed his example by building a rectangular capital in which the main streets paralleled the rectilinear walls.[23]

It is quite likely that the Western practice of building gridded towns and cities has its roots in the bold experiments of the Babylonian and Assyrian kings of ancient Mesopotamia. Yet as is often the case with the ancient civilizations of the Near East, it is hard to trace a direct line of transmission between their cultural and artistic accomplishments and the later Western tradition. For that, as for the origin of so many Western practices, we need to turn to the ancient Greeks, and in this case to the work of Hippodamus of Miletus, the first known urban planner. After his ancient home city of Miletus in Asia Minor was destroyed by the Persians in 479 BCE, Hippodamus had it rebuilt as a regular grid, and some years later the Athenians hired his services to build their new port of Piraeus along similar lines. It is quite likely that Hippodamus's grids were inspired by earlier examples, whether in Mesopotamia or among his Greek predecessors. We do not know for sure. The crucial thing is that unlike any previous instances of the grid, Hippodamus's design had an enormous and almost instantaneous impact. Soon gridded cities began to appear throughout the Greek world, from the shores of the Black Sea to the western coast of Italy. It became the standard design for any new Greek colony founded in the Mediterranean.

Before the grid would advance westward, it first turned south and east, following in the wake of Alexander's conquering armies. Wherever the Macedonian king established client kingdoms or settled his veterans on land wrested from his enemies, he established colonies inspired by the latest Greek practices. As a result, a string of Hellenistic cities founded by Alexander and his successors soon stretched from Asia Minor to Egypt and eastward through central Asia to the border of India. As a general practice, these cities were designed in a grid pattern, which thereby returned to the Indus Valley for the first time since the days of Mohenjo-Daro.

But it was the cities of Magna Graecia in southern Italy that would have the greatest impact on the development of the grid. By the early third century BCE, Rome was firmly in control of all of southern Italy, and the formerly independent Greek cities had been reduced to the status of subservient allies and dependents. To the conquering Romans, the brilliance and sophistication of Greek culture came as a shock, and while some in the elite felt irresistibly drawn to it, others worked hard to keep its influence in check. It was all very well, they argued, for effete and overly cultured Greeks to concern themselves with philosophy, poetry, art, and architecture. But if the austere and manly Romans followed suit, they would soon lose the martial virtues that had made their city great. Then Rome, just like the once-proud cities of Magna Graecia, would soon lie prostrate before its enemies.

This conservative opposition, led by aristocrats such as Cato the Censor, proved powerful and sustained enough to delay the assimilation of Greek culture into Roman life by two centuries. But in the first century BCE, as dozens of Roman legions marched from one end of the Mediterranean world to the other and clashed in a seemingly endless succession of civil wars, resistance to Hellenization finally crumbled. Many in the Roman elite, disillusioned by the destruction all around them, sought refuge in the beauty and sophistication of Greek learning. It was the beginning of a golden age of Roman culture that would see the flowering of uniquely Roman strains of philosophy, history, and literature. It was also the time when the Romans finally adopted the Greek practices of city planning, which they had known for centuries.

Rome itself was not built on a grid. A medium-sized city that in the space of a few generations became the greatest metropolis of the ancient world, it grew organically, without a central plan, into a massive, near-impenetrable maze of streets and alleys. But when, in the final decades of the republic and for nearly two centuries under the empire (roughly from 100 BCE to 200 CE), the Romans founded new cities throughout their vast domains, they took a different approach. Some of these cities, such as Pavia and Verona, were founded to settle the thousands of veterans who had manned the Roman legions in civil and foreign wars. Others, such as Cologne in Germany and Scythopolis in modern Israel, were founded to defend the empire's borders. Still others, such as Aelia Capitolina (also known as Jerusalem), were built on the ruins of former enemy cities to impress upon a hostile population their subjugated condition. All, however, were built in accordance with a simple and standard grid plan, inspired perhaps by the design of a Roman army camp.

The new Roman cities were walled squares, roughly oriented to the points of the compass and divided into four equal quarters by two main streets: the Cardo Maximus, running from north to south, and the Decumanus Maximus, running from east to west. The main government buildings were located at the center of town where the Cardo and Decumanus met, and the lesser streets ran parallel to the two at more or less regular distances, creating a gridded pattern of standard squares. It was, to be sure, a practical arrangement for a new frontier city, in which lots could be easily divided among the colonists and defenders could be rushed quickly from one section of the wall to another in case of attack. But it was also much more: Roman power and principles written into the urban geography of its provincial towns. The Romans believed in rational order—legal, administrative, and military—and never doubted that it was the key to their unprecedented power and success. The strict, unvarying, standardized grid of Rome's colonial cities expressed these core characteristics like nothing else. To enter within the walls of Cologne, Verona, or Aelia Capitolina was to feel the weight of Roman rule.[24]

By the third and fourth centuries CE, as the power of Rome slowly

waned in western Europe, the standard grid plan was relaxed, modified, and eventually disappeared altogether. This was due, in part, to
the weakening of central authority in the empire: when private interests prevailed over directives from the capital, it was well-nigh inevitable that changes to the standard design would be made to satisfy
the parochial interests of local power brokers. Another factor was the
sharp decline in the population of the empire, which left many cities
partially empty and many towns abandoned. In such a situation the
local populace would make their own arrangements in the city rather
than adhere to an old plan that no longer suited their needs. Finally,
when the last Roman emperor was deposed in 476 CE, urban life in the
West went into a long eclipse that lasted more than half a millennium.
With the population and economy overwhelmingly rural and the land
subject to the depredations of nomads and seafarers, even formerly
great cities became little more than country market towns. The landed
aristocracy, which monopolized military and political power, resided
in their own castles in the countryside and viewed the towns as little
more than sources of income. Under those conditions, there was no
authority to enforce a standard street layout, and in any case, no interest in it either.

It was only in the High Middle Ages, roughly between 1100 and
1300 CE, that the grid plan reemerged in western Europe. By this time
trade and the urban economy were once more on the rise, the population was growing rapidly, and Europeans were busy turning forests
and marshes into farmland and pushing beyond old boundaries into
new territory. Spurred by the new dynamism, hundreds of new cities
were founded in those centuries throughout the region, from Spain to
Germany and from Italy to England. Some of those cities, taking their
cue from surviving Roman towns, adopted a form of the grid as their
overall street plan. Since the implementation of a uniform street plan
still required an effective central authority, grid cities emerged where
mighty kings or powerful landed aristocrats held sway. In France, for
example, both king and nobles founded new towns known as "bastides" that doubled as military outposts on the borders of their territories or in potentially restive provinces. Much the same was true

in England, where the Plantagenet kings, who had vast holdings in France, extended the pattern into the English countryside. Though hardly as uniform as the Roman foundations a thousand years before, these new medieval towns also followed a common pattern: enclosed within square fortifications, they were divided into more or less regular blocks by streets that ran parallel to the defensive walls.[25]

The period of rapid expansion came to a sudden and horrific end with the Black Death, which between 1347 and 1349 wiped out a third of the population of western Europe. At a stroke, the demographic pressures in both the cities and the countryside were removed, and with them the drive to found new towns. It would be two hundred years before the population recovered, and even when it began to grow once again in the sixteenth century, few new cities were founded. Much of this excess humanity simply settled in existing ones, turning major cities such as Paris, London, Florence, and Antwerp into giant metropolises. As for more adventurous souls who wished to settle new and untrammeled lands, they now had a whole new world open before them: the newly discovered continent called America.

Civilization in the Wilderness

The first European settlement in the Americas was founded by Christopher Columbus on December 25, 1492. The admiral's flagship, the *Santa Maria*, had foundered on a sandbank off the island of Hispaniola, and he decided to use her timbers to build a fortified settlement. There, he proposed, some of his men would winter, trade with the native people, and search for gold while awaiting his return the following year. He called it Navidad (Christmas), in honor of the holy day of its founding. But whereas the date seemed auspicious, this first colony's fortunes proved anything but. Columbus did return as promised the following year, only to find Navidad in ruins and the thirty-nine men he left behind missing, the apparent victims of an attack by their Indian neighbors. The first European attempt at settling the New World had ended in utter disaster.

Had things ended there, America today might still have been ruled by the descendants of the Aztecs, the Inca, the Maya, and other native peoples. But as we know, things did not end there: Columbus's flotilla was followed by many others, and the failed town of Navidad was soon followed by more enduring settlements. By the early 1500s, Spanish towns and encomiendas (plantations) had taken root in the New World, first on the Caribbean islands and later on the mainland as well. In the following decades, Spanish adventurers would explore the New World, from the highlands of Mexico to the mountains of Peru and from the Mississippi Valley to the Amazon rainforest. Wherever they could, they planted towns, lonely outposts of European life on an unimaginably vast continent.

Early on, the towns varied—some were built with a precise street plan in mind; others were allowed to grow organically of their own accord. By the middle of the century, however, Spanish authorities were issuing instructions regarding the foundation of new settlements, and the towns gradually took on a more standard form. Finally, in 1573, King Philip II issued a royal proclamation that summarized and codified these instructions. Known as the Laws of the Indies, it specified precisely how a settlement should be built and structured throughout the Americas. And since the laws were never revoked, they would govern the foundation of new towns throughout Spanish America right until the demise of the Iberian empires in the early nineteenth century.

According to the Laws of the Indies, each Spanish town had to be built around a large central plaza of carefully defined proportions: "It shall not be smaller than two hundred feet wide and three hundred feet long nor larger than eight hundred feet long and three hundred feet wide. A well-proportioned medium size plaza is one six hundred feet long and four hundred feet wide."[26] The plaza, furthermore, should not be laid out randomly but in such a way that its corners point to the four points of the compass. This alignment, the edict explains, would prevent unhealthy exposure to "the four principal winds." All the principal buildings of the town would be positioned in carefully specified locations around this central plaza: the main church, of course, which is the spiritual and social heart of a Spanish town, but also the town hall, the customs house, the arsenal, and the hospital. Major streets, straight and broad, would emerge from the center of each side of the plaza and at right angles to it, and two narrower streets, contiguous to the sides of the plaza, would emerge from each of its corners. Additional streets, further removed from the plaza, would run parallel to the main ones, thereby completing the grid. The resulting pattern for a Laws of the Indies town is of a rectangular central plaza surrounded by a checkerboard of rectangular blocks, bounded by straight-arrow streets that intersect at right angles.[27]

From our perch in the twenty-first century, it seems surprising that the Laws of the Indies would have much of an impact on the actual settlement of the New World. After all, the laws were a dry legal document drafted by court lawyers who had never set foot in America and had little knowledge of conditions there. They were then issued by a king from his capital across the ocean, at a time when even the simplest communication between the colonies and the mother country required months of hazardous travel in each direction, with no guarantee that the message would ever reach its destination. Under these conditions, one might wonder why Spanish settlers in the Americas would pay much heed to what the courtiers in Madrid had to say. As the people on the ground, they were far better equipped than King Philip's lawyers to determine the best layout of their town. Furthermore, being so far removed from the royal capital, both in terms of distance and of time, they would be unlikely to suffer any consequences for their intransigence. The Laws of the Indies could easily have become a dead letter before they even reached American shores.

And yet they did not. For despite everything—despite the fact that urban plans hatched across the ocean were often ill suited to conditions on the ground, despite the remoteness of the settlements from the royal capital, and despite the notoriously inefficient imperial bureaucracy—for two centuries the cities and towns of Spanish America were built as the Laws of the Indies prescribed. Some accommodations were, inevitably, made for local conditions, and some variations were even accounted for in the laws themselves. The plan for coastal cities, for example, was somewhat different from the plan

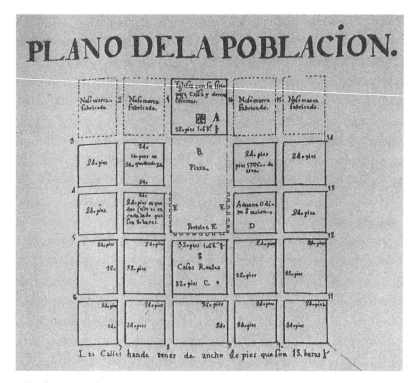

Plan from 1730 for San Fernando de Béxar, modern San Antonio, built according to the Laws of the Indies. "Mapa de plano y perfil de la Poblacion que se ha de hacer la qual estaarreglada a las leyes Reales de Indias," #01547, map collection, Archives and Information Services Division, Texas State Library and Archives Commission.

for inland sites. But these were only minor variations. From the city of Mendoza in Argentina (founded 1561) to San Fernando de Béxar (San Antonio) in Texas (founded 1730) and even to the Pueblo de Nuestra Señora Reina de Los Angeles (Los Angeles) in California (founded 1781), the towns of Spanish America were built as a regular rectilinear grid surrounding a large central plaza.[28]

The reason for this is that the cities of Spanish America were not founded simply to accommodate the needs of the local settlers in a remote and often hostile environment. Much like the Roman colonies before them, they too strove for something higher: a physical expression of the ideals and the power of the empire that had built them. In the Spanish Empire, in principle if not always in practice, all power flowed irresistibly from the king at the top through his underlings and down to the lowest peasant. The power of the king disseminated to every corner of his domains, bringing with it rational order, a clear social hierarchy, and civic peace. It was an idealized view of an often-turbulent reality, but this did not diminish its power. The Spanish colonists, no matter how far removed they were from the centers of royal power, always saw themselves as the bearers of these imperial ideals into new and uncharted lands. And so, when they settled down to build a town or city, it was imperative that they proceed as true Spaniards did by obeying the decrees of their far-off sovereign. To do otherwise would not be merely a matter of circumventing a particular law; it would be a betrayal of who they were and of the reason they traveled to this distant land.

By obeying the royal decree in territories where no enforcement was possible, the colonists were demonstrating that the king ruled and that the mighty Spanish Empire extended even to their remote locale. The actual town that they built, as laid out in the Laws of the Indies, only reinforced the message. The grand central plaza was incontestably the seat of power in each town, where the king's representative resided, sanctioned by God through the holy church. The other institutions represented around the plaza, such as the customs house and hospital, served as a testament to the rationality and benevolence of the royal administration. Beyond the center was a rational and orderly rectilinear world, at peace under the sway of the king's power, which emanated from the plaza. The Laws of the Indies towns of Spanish America were strictly hierarchical, orderly, rational, and peaceful. To those who came within their gates, whether Spaniards or local Indians, they were enclaves of Spanish ideals and power in a vast and alien continent.

In establishing rectilinear towns in the Americas, the kings of Spain and the administrators who served them were transplanting the ancient European tradition of the urban grid onto the soil of the New World. This was not a coincidence but a conscious choice to emulate the classical world of Greece and Rome. In fact, most of the prescriptions of the Laws of the Indies have precedents dating back to antiquity, and many passages are direct paraphrases from the writings of Vitruvius, the first-century Roman architect and engineer who was the highest authority on both domestic and urban planning.[29] It should come as no surprise, therefore, that the Spanish towns founded in the Americas between 1500 and 1800 CE and the Roman towns founded in the imperial provinces a millennia and half earlier were strikingly similar.

Both the Spanish and Roman settlements were founded on the borderlands of empire, far from their respective capitals and often surrounded by hostile locals. Both were built as commercial and administrative centers but also served as imperial military outposts. Both gave primacy of place to administrative and civic buildings, though the Spaniards, who built their towns around a massive central plaza, emphasized them even more than the Romans. Both had clear boundaries and were, most often, surrounded by square or rectangular walls. And both, finally, opted for a uniform rectilinear grid as their street plan.

As a result, the Roman and Spanish towns served their respective empires in similar ways: To the colonists who inhabited them, they were beacons of "civilization" in the wilderness, islands of imperial law, order, and peace in an ocean of chaos that forever threatened to devour them. To the locals who lived in the surrounding region, they were reminders of the power of an empire that sought to subjugate them, inspiring fear, sometimes awe, and often hatred. The Spaniards sought to enhance this effect by forbidding the local Indians from entering the town until its fortifications and buildings were complete. The reason, the Laws of the Indies specify, is that when the Indians see the completed town, "they will consequently fear the Spaniards so much that they will not dare to offend them and will respect them and desire their friendship."[30]

And that, indeed, was the key function of the classical Western grid from the time of Hippodamus to the Laws of the Indies: to distinguish between civilization and barbarism. The rectangular walls surrounding the towns marked a clear borderline between the world of the city and the uncivilized world beyond. The clear boundaries also meant that the town had a well-defined center and that the city space was therefore hierarchical, with the power and prestige flowing from the center toward the town's periphery. And most importantly, the arrow-straight lines of the streets, the precise right angles of their intersections, and the regular uniform blocks that made up the city spoke of a rational, orderly, predictable, and controllable world. All of this drew a sharp contrast between the walled city and the world outside: inside was order, reason, hierarchy, and ultimately peace. Outside—at least in the eyes of the town dwellers—was chaos, violence, and disorder.

Eastern City, Western Plains

The conviction that a regular gridded town plan kept barbarism and wilderness at bay was not unique to the Spanish settlers of America. The British settlers who followed in their footsteps had a similar notion as they established towns and cities on the Atlantic Seaboard in the seventeenth and eighteenth centuries. The largest and most important was Philadelphia, founded in the early 1680s by William Penn as the capital of his new colony of Pennsylvania. Though a relative latecomer among the chief cities of British America, Philadelphia grew quickly, overtaking Boston as the most populous city in the colonies by the mid-1700s before being itself overtaken by New York at the end of the century. It was also, famously, the meeting place of the Continental Congress, the first capital of the United States, and the most frequent capital before Congress's move to Washington, DC, in 1800.

The rapid growth of Philadelphia was due in part to the ambitions of William Penn, who in 1681 had announced his intention to build "a large town or city" in Pennsylvania. By 1683, he had selected a location for the city, which was to stretch between the Delaware and Schuylkill Rivers. He also had the site surveyed, with streets and lots marked, and had a map printed of the city's projected outlines. As Captain Thomas Holme, Penn's chief surveyor, described it, "The City of Philadelphia,

Now extends in length, from River to River, two miles, and in breadth near one mile." These dimensions alone would easily make the city the largest by far in British America. He continued to describe the layout of the streets: "The City . . . consists of a large Front-street to each River, and a High-street (near the middle) from Front (or River) to Front, of one hundred foot broad, and a Broad-street in the middle of the City, from side to side, of the like breadth." Other streets would run parallel to the two main axes, "eight Streets (besides the High-street), that run from Front to Front, and twenty Streets, (besides the Broad-street) that run cross the City, from side to side." Each of these streets, according to Holme, would be of "Fifty foot breadth," or half the width of the two major streets. A ten-acre square would grace the center of the city, with "a Meeting-House, Assembly or State-House, Market-House, School-House," and "other buildings for Publick Concerns" at each corner.[31]

William Penn's ideals of government by consent and religious freedom could hardly be more different from the Spanish kings' insistence on rule by decree, strict hierarchy, and religious conformity. In some ways, his city reflects these different priorities. Philadelphia's location on the banks of two rivers suggests a city built for commerce rather than a military and governmental outpost. The same is true for the size of the population envisioned in Philadelphia's plan, which was much larger than that of a new Spanish town, and for the fact that unlike most Spanish outposts, Penn's city was not walled and could be entered from any direction. And although Philadelphia's ten-acre central square was larger than the plazas prescribed in the Laws of the Indies, it was tiny when compared to the size of the city as a whole. In most Spanish establishments, the central plaza took up at least a tenth of the area of the new town; in Philadelphia, it was less than one hundredth.

All of these point to critical differences between the Spanish grid towns and Penn's ambitious plan. The purpose of the Laws of the Indies settlements, from Argentina to California, was primarily to project imperial power; the purpose of the new city of Philadelphia was to encourage commerce and prosperity in a benign environment of religious toleration. And yet, despite the contrast in ideology and purpose, the ground plan for Philadelphia was essentially a blowup of a Laws of the Indies town. The sharp rectangular boundaries of the town are there in Penn's city as in the Spanish towns, as is the central plaza where all the main civic institutions are located. And most strikingly, both plans are structured as a strict rectilinear grid: in each, a set of parallel, regularly spaced streets intersects another set of parallel, regularly spaced streets at precise right angles, creating a landscape of regular rectangular blocks. The fact that a street plan designed to enhance the absolute rule of the kings of Spain could be used—with relatively minor tinkering—in a city founded by a radical pacifist Quaker points to the enormous versatility of the grid as a city plan.

Does this then mean that the urban grid is endlessly flexible, that it is merely a convenient arrangement that can be used for any purpose? And that consequently the grid plan, in and of itself, stands for nothing? Far from it. For wherever and whenever the grid was used, through thousands of years and across continents, there are a few unshakable core principles that it stood for. Whether in Aelia Capitolina, San Antonio, or Philadelphia, the grid plan proclaimed rationality, order, and peace. The meaning assigned to these could, undoubtedly, vary considerably, as the practical rationality of the Romans differed from the pious rationality of the conquistadores; the relatively egalitarian order of the citizens of a Greek city contrasts sharply with the hierarchies of the Spanish Empire; and the peace imposed by Roman legions on their former enemies is very different from the peace acquired through flourishing commerce. Yet despite these differences, wherever the grid plan was used we know that rationality was valued and a stable social and political order prevailed, as did a degree of peace. We know this even about a city such as Mohenjo-Daro, about which we know almost nothing else. We know this because without rationality, order, and a degree of peace, there would be no city grid.

What the classical city grid does is separate the world into two contrasting realms. Within the rectangular city walls is a rational world of order and peace—or, in other words, civilization. Outside the walls, all is chaos, disorder, and strife—or, in other words, wilderness. The purpose of the grid is to establish the former and contrast it with the latter and to set a clear hierarchy between them: civilization, the world inside the city's boundaries, is immeasurably superior to the wilderness outside the city gates. This was true of Roman frontier colonies and Spanish American settlements. It was true as well of the grid cities of British America, including the largest among them, Philadelphia.

The cityscape of a classical urban grid, such as the one found in ancient Cologne or in Philadelphia, is unquestionably geometrical. The arrow-straight streets, the regular spacing between them, the precise right angles, and the (usually) rectangular boundaries all testify to the planners' intent that their city be experienced as a geometrical world. Much the same, though on a far larger scale, can be said of the Great Western Grid, conceived by Jefferson and implemented by a long succession of surveyors over a century and a half. In both cases, an irregular natural terrain was transformed into an artificial checkerboard landscape of straight lines, right angles, and regular rectangular blocks.

Yet there is also a core difference between the classical urban grid and the Jeffersonian one. The urban grid drew a distinction between civilization within and wilderness without. It divided the world into two unequal parts, turning one part into human civilized space but leaving most of the world untouched. Geometrical order, it suggested, belongs to human culture and to it alone, whereas natural spaces are chaotic, irregular, and distinctly ungeometrical. The Jeffersonian grid of the western United States was far more ambitious: instead of dividing space into the human and the natural, the geometrical and the disordered, it transformed space everywhere, whether occupied by humans, wild beasts, or nothing at all. The western survey did not require human settlers or even human presence in order to reconfigure the land: it turned forests, rivers, mountains, and prairies—all of them—into a regular geometrical grid. The classical grid suggested that people can create an artificial geometrical space that would be suitable for civilization; the Jeffersonian grid implied that space itself, everywhere and always, is by its very nature geometrical.

The two types of grids, the classical and the Jeffersonian, appear remarkably similar, and it is easy to see how surveyors trained in one would be inspired to try their skills in the other. For surely Hutchins, Mansfield, and their fellow surveyors were well familiar with Philadelphia and other gridded towns when they set out to carve the American West into regular squares. Yet at their core, the two draw on profoundly different conceptions of geometry. The classical grid is fundamentally Euclidean: As in any Euclidean construction, all that exists is what has been carefully and deliberately constructed. Every Euclidean proof begins by positing a point, a line, or a figure and then adding additional elements as the demonstration requires. Outside of this constructed figure, in a Euclidean universe, there is nothing—not even empty space. Exactly the same is true of the towns and cities built according to the classical grid: the towns themselves are geometrical constructs, built to reflect the needs and ideals of their founders. Beyond them there is nothing, just a wilderness unsusceptible to geometrical control.

The Jeffersonian grid, in contrast, is not rooted in Euclidean geometry but is a physical manifestation of the absolute space of analytic geometry. In analytic geometry, mathematical space exists in itself, even if it contains absolutely nothing—not a figure, a curve, or anything else. It is defined by a Cartesian coordinate system, which marks every point within it as unique, even if there is nothing occupying it. It is in this absolute space that mathematical objects such as lines and curves can then be defined, created, and manipulated. But absolute coordinate space preexists them all.

The Jeffersonian grid is therefore fundamentally different from the city grids of Spanish and British America. The latter were constructed with boundaries, a center, and an internal order and hierarchy. Jefferson's Great Western Grid, in contrast, possesses none of those. It is a Cartesian coordinate system that defines the western United States as a vast and empty space in which anything can be created—just as in analytic geometry. Whereas the Euclidean gridded cities presented particular orders and ideas, the analytic Jeffersonian grid presented only a blank slate—an empty land of opportunity in which anything was possible.

All of which is to say that gridded cities are as old as the most ancient civilizations, or very nearly so; that they appeared in different societies and in different cultures, and in the West they firmly took root with the Greeks and then the Romans; that they were brought to the Americas by the Spaniards, who used a standard template for their towns; and that they were later adopted by other American colonial societies, such as the British, who built numerous grid towns, including the largest and most ambitious of all—Philadelphia. And yet all of those multitudes of cities, founded over several millennia, were built as classical Euclidean grids—geometrical islands of reason and civilization in an ocean of uncontrolled chaos. It was not until the nineteenth century that a city was built on the principles of the Great Western Grid, not to enclose itself from the surrounding world but to reshape it. This city was New York.[32]

5

Making the Greatest Grid

The City of Nasty Streets

The island of Manhattan was not meant to be the thick urban grid of asphalt and cement we know today. When Henry Hudson steered his ship *Half Moon* into what would become New York Harbor in September 1609, he could scarcely have imagined what lay in store for the elongated island at the mouth of the river that would bear his name. Before him was a land of thick forests punctuated by green meadows, spread over steep hills and valleys, and dissected by numerous creeks and rivers. To Hudson and his crew the island may have appeared pristine, but it was in fact settled by Lenape Indians who had lived there for centuries, if not millennia. When Hudson sent a boat to survey Manhattan's jagged coastline, the locals made their displeasure known by killing his second mate with an arrow to the neck.[1]

Though himself an Englishman, Hudson was sailing in the service of the Dutch East India Company, which was looking for a northwest passage through America to the Pacific. Hudson, needless to say, did not find the desired passage.[2] But he did report back that the yet-to-be-named Hudson River opened up promising grounds for beaver trapping, at a time when demand for beaver pelts was growing fast among upper-class Europeans. In 1624 the Dutch founded a trading post on the Hudson River at the site of modern Albany and named it, appropriately, Beverwyck.[3] A year later they built a fort on the southern tip of Manhattan to guard against European rivals and serve as a gateway to the North American interior. When a trading settlement grew up around the fort, it became known as New Amsterdam.

New Amsterdam existed for all of four decades, but during that time it became a bustling town with a population that rose quickly from a few hundred to around 1,500 residents.[4] A military as well as commercial outpost, it was a fortified settlement, protected to the south by Fort Amsterdam and to the north by a wall that stretched the width of the island, along the line of present-day Wall Street. All this proved to be of little use in August 1664, when a flotilla of English warships dropped anchor in the harbor and demanded the town's surrender. After vacillating for several days but seeing no way out, Director General Peter Stuyvesant, the top Dutch official, capitulated. He handed over the keys to the town without firing a shot, and Dutch New Amsterdam became British New York.

It did not take long for New York to take its place as the fastest-growing and most dynamic city on the Eastern Seaboard of North America. Located at the mouth of a great navigable river and possessed of a superb natural harbor unmatched by any of its rivals, it soon became a top destination for European trade and a leading commercial hub for the colonies. By the end of the seventeenth century, the town's population had more than tripled, and by 1775, as war broke out, New York had become a city of twenty-five thousand and had overtaken Philadelphia as the largest metropolis of British America.

As New York grew by leaps and bounds, it easily outstripped the ability of the city magistrates to manage and regulate its rapid expansion. The city charter of 1686, named after provincial governor Thomas Dongan, gave the city council, in principle, "full power . . . to establish, appoint, order, and direct the establishing, making, laying out, ordering, amending, and repairing of all streets, lanes, alleys [and] highways . . . in and throughout the . . . city of New York and Manhattan island."[5] It was an ambitious mandate: at a time when New

York City had not yet expanded beyond what is now the Financial District, the Dongan Charter had already given it jurisdiction over the entire island and authority to lay down streets and roads—even at the expense of local property owners. But impressive though it appeared on paper, the charter provided the city with a weak municipal government and few tools with which to enforce its supposedly vast authority. The Common Council, the city's highest governing body, was elected annually and suffered from substantial turnover from year to year, which discouraged the formation of any long-term plan. Whatever plans were nonetheless hatched were rarely implemented because the council had no permanent staff to follow up and enforce its decisions. Subsequent charters in 1708 and 1731 did not improve things, and the city officeholders showed little inclination to make use of even the limited means they did have at their disposal. As a result, for nearly a century and a half after it became British, the city of New York grew and expanded without any master plan or central control.

"Their streets are nasty and unregarded," declared Dr. Benjamin Bullivant, who visited New York in 1697 and found it inferior in all respects to his home city of Boston.[6] It was a widely shared complaint, and it was, no doubt, justified, for compared with rival colonial cities New York was indeed a mess. Philadelphia, a city younger than New York, had a visionary master plan, which, while not always followed, nevertheless guided the city's growth and expansion for centuries after its founding. Boston, to be sure, lacked a master plan but made up for it with a strong communal spirit, born of its root in the tight-knit puritan communities that first settled Massachusetts. Woe befall the man who selfishly put his own material interests before those of his fellow Bostonians.

New York possessed neither the Philadelphian master plan nor any evidence of a Bostonian civic spirit. It had nothing, in fact, to restrain the free play of private interest in the development of its own urban landscape. With no urban plan and no administrative bureaucracy, the job of laying down streets in New York fell to the multitude of property holders who owned the land on which the city stood. Not unnaturally, they preferred streets that served their properties and the houses they owned, and in particular such streets likely to increase their property value. And so, with little interference from city authorities, New York grew into a baroque maze of streets and alleys that may have served the immediate interests of local landowners but hardly anyone else.

A painting by an unknown artist of the Five Points neighborhood in 1827, though dating from a later time, gives a good sense of what early New York must have been like. Respectable ladies and gentlemen intermingle with crowds of men and women of the lower classes in crooked and muddy alleys where horses, pigs, and sheep roam freely. It is a scene not without charm and one that might appeal to modern-day advocates of urban renewal as well as to our multicultural sensibilities (some of the people depicted in the street scene are clearly African American). But whereas modern urban planners seek the appearance of vibrant disorder in what is in truth a meticulously planned environment, the chaos in early New York's irregular winding streets was all too real.[7]

But just as New York's disorder derived from private initiatives, so did the first attempts to impose a semblance of regularity on its street plan. Already in the 1750s Trinity Church, one of the largest landholders in the city, had laid out several blocks between Broadway and the Hudson River in the shape of rectangular blocks. At their center was King's College, the church's school of higher learning, later renamed Columbia University. A decade later the wealthy DeLancey family carved up their lands in a similar manner in what is now the Lower East Side, creating the first numbered street system in New York. When the loyalist DeLanceys were forced to flee, after the revolution but long before they built up their prospective grid, the city took over the lands and kept much of their plan. These private street grids (another was created by the Bayard family in the 1780s) were small and oriented along different axes, creating a patchwork that could not be combined into a citywide plan. Nevertheless, they suggest that from the beginning a regular rectilinear pattern was most attractive to those who tried to impose order on the chaotic city.

The Revolutionary War was disastrous for New York, more so than

Five Points, artist unknown, 1827. From the Metropolitan Museum.

for any other major city in the colonies. The Declaration of Independence in July 1776 made New York the largest city of the newborn United States, but the celebrations did not last long. In August General William Howe's redcoats defeated Washington's colonials in the Battle of Brooklyn, and, over the next several months, they methodically drove the American army up Manhattan Island and into New Jersey. On September 15, even as thousands of patriots were fleeing before the British advance, Howe entered New York and set up his headquarters in the city. It would be home to the British high command for the duration of the war.

New York was already half-abandoned when a fire broke out on September 21, which left a third of the city in ashes. Whether the fire was intentional—started by Washington's agents or vengeful loyalists—or

accidental was never determined, but it set the tone for the bitter occupation that followed. For the next seven years, even as its western
neighborhoods lay in ruins and loyalist refugees camped out in its
squares, New York suffocated under martial law. When the triumphant patriots returned in 1783, the loyalists fled, including some of
the city's oldest and wealthiest families, the DeLanceys among them.
By the time the last British soldiers had departed in November 1783,
the city had been reduced to half its former population.

But while the war was ruinous for New York, it could also have
been an opportunity: with the city depopulated and a sizable part of
it in ruins, the city fathers had a unique chance to start anew and put
in place plans for orderly urban growth. For the most part, however,
they let the opportunity slip by. It took several years for the New York
economy and population to bounce back to its prewar levels, but by
1790 it was once again the largest, most dynamic, and fastest-growing
metropolis in the United States. It was an impressive recovery by any
measure, made all the more so by the fact that it was accomplished
with no plan and little guidance from municipal authorities. Once
again it was not the Common Council but individual landowners who
planned and laid out streets as they saw fit. And as commerce thrived
and the city's population ballooned to sixty thousand by the turn of
the century, the tangle of streets that was New York grew ever more
convoluted.

This does not mean that city leaders were entirely idle in the matter of surveying its extensive lands. Even though much of the land of
Manhattan, granted to New York in the Dongan Charter of 1686, had
been sold off to private owners over the course of a century, the city
still held title to several hundred acres known as the Common Lands
in the middle of the island. The lands were rugged and located far to
the north of New York's boundaries, but they were likely the city's
most valuable possessions at a time when it was struggling to recover
from the war's devastation. In 1785, when Congress sent Thomas
Hutchins to the Ohio Territory to survey and divide the land so that
it could be sold to prospective settlers, New York's Common Council
decided to do the same for its own land holdings. To accomplish this,

the councillors hired Polish immigrant and accomplished surveyor Casimir Goerck and charged him with mapping the Common Lands and dividing them into five-acre plots. Goerck completed the job within a year, but sales proved disappointing. And so, ten years later, the council tried again, once more hiring Goerck to conduct a more thorough and accurate survey. By the time Goerck was done, he had laid out three broad and parallel north–south roads, 920 feet apart, intersecting every 200 feet by narrower east–west streets, thereby creating a regular rectilinear grid of standard plots.[8]

Goerck and the councillors who hired him could not know it, but his survey was the first imprint of the future New York grid on the island of Manhattan. The three north–south roads would ultimately mark the paths for 4th, 5th, and 6th Avenues. The narrower cross streets, most of which only existed in Goerck's maps and were never actually laid out, would provide the template for the regular east–west streets of the New York grid, which would retain the same width and distance apart. The earliest foreshadowing of the Manhattan grid, in other words, was not meant as an urban plan at all but as a rural agricultural one.

This is significant. Philadelphia, like other early gridded cities in North America, drew its inspiration from the long tradition of rectilinear planning going back to the Laws of the Indies towns, Roman colonies, and beyond. Their grids were unmistakably urban in nature, designed to emphasize rationality and order and to mark them as civilized spaces in contrast to the wilderness beyond. But the New York grid took its inspiration from a completely different source—the great rectilinear survey of the western United States, which unlike the urban grids did not have a boundary or a center. The western grid presented no particular hierarchy or order and was not at all concerned with contrasting the spaces inside and outside the grid. Its purpose was to transform all of space everywhere, overlay it with a Cartesian grid, and mark it as graph paper—a mathematical blank slate readied for human action. The same, as we shall see, would be true of the Manhattan grid: it began as an agricultural land scheme modeled on the rectilinear western survey that was launched at the same time;

it would end by transforming the entire island of Manhattan into an abstract mathematical landscape in which anything was possible.

New York as It Will Be

It would be decades before rising real estate values would justify the Common Council's high hopes for the Common Lands, but in the meantime it turned its attention to more pressing issues. In 1797, for the very first time, the council commissioned an official accurate map of the city. Goerck, fresh from his survey of the Common Lands and eager for more publicly funded work, immediately submitted a proposal. His map, he promised the councillors, would be "general and accurate" and cover the entire area in which "the City is laid out into Streets and Lots." Goerck by this time had a solid track record of work for the city, and he probably would have received the commission were it not for the appearance of a rival with equally strong, if not superior, credentials.

Joseph François Mangin (born 1758) was a French military engineer who in 1789 was sent to Saint-Domingue in the Caribbean (modern Haiti) to conduct a complete survey of the colony. Saint-Domingue was France's most valuable overseas possession, thanks to its thriving sugar plantations worked by African slave labor, and at first it seemed that wealth and distance would shield the colony from the turmoil back home. But in 1793 the revolution arrived on the island as the slaves rose up against their masters, leading to a long and bloody war that would end eleven years later with Haiti's independence. Mangin was forced to flee with his family, but for a committed royalist like himself, returning to revolutionary France at the height of terror was not an option. Instead he headed for New York, where he arrived early the next year seeking refuge and employment. Before long, he joined the circle around an American statesman and New Yorker who also had Caribbean roots, royalist sympathies, and an aversion to revolutionary France. His name was Alexander Hamilton.

In New York, Mangin found success. Within three years of his arrival, he had already been commissioned by the Common Council to redesign the fortifications of the city's port, to design the city's first purpose-built theater, and to draw the ground plans for the first state prison, up island in the village of Greenwich. In later years he would go on to design several churches, including the first St. Patrick's Cathedral and, most famously, New York's third city hall, which still stands today. Mangin's uncanny ability to land city contracts was due in no small part to his proven skill as a surveyor and engineer, but he also possessed a penchant for unapologetic self-promotion that may have impressed city officials. "The minor projects that I have so far executed in Newyork cannot allow you, sire, to fully judge of my abilities," he complained to Hamilton in 1799. His success in all his undertakings, he added with a pomposity that rings clear through the centuries, was due to the fact that he combined his vast knowledge "with a Theory which does not allow me to be wrong."[9] Finally, when evaluating his success, one cannot discount the influence of his patron. A protégé of Hamilton was bound to do well in New York.

When in 1797 New York officials asked for proposals for a map of their city, it first appeared they would have to choose between the solid and well-regarded Goerck and the brash and well-connected Mangin. Instead, the two joined forces and presented a proposal for the most detailed map ever made of the city, one that would mark every house, lot, and square and note its precise elevation. The dazzled councillors did not hesitate; they accepted the proposal and paid the $3,000 Goerck and Mangin had requested. For the next year, the two crisscrossed the city with their surveying instruments, making measurements and taking notes. But the summer of 1798 witnessed one of the deadliest outbreaks of yellow fever to which cramped, chaotic, and—in truth—filthy New York was prone. Goerck caught the disease and died, leaving Mangin to complete the map on his own. On April 10, 1799, the Frenchman presented the Common Council with what he called *The New Map of the City of New York*.

Yet something was not right. Goerck and Mangin had been commissioned to produce a map of the city as it then existed, but that was not what Mangin presented to the council. His map not only altered the actual layout of city streets but also tinkered with the physical outline of the island. Most startlingly, it more than doubled the area of the city, charting out avenues, streets, and entire neighborhoods in locations where at the time cows grazed in green meadows among country roads. Why Mangin deviated so far from his mandate is not clear. Perhaps, as he wrote to Hamilton, he felt that the job of a city surveyor was beneath him, and he wished to show his full talents by taking on a more ambitious challenge. Perhaps he was envious of his countryman Pierre L'Enfant, whose own redesign of New York's city hall in 1788 was followed by a commission to design an entire city— and the national capital at that. What is clear is that at some point Mangin decided to be not a city mapmaker but a city planner who would outline and chart the future growth of the city. He would be New York's L'Enfant.

The first hint of Mangin's unorthodox plans came in December 1798 when fellow surveyor Charles Loss, who was working on a redesign of the harbor, asked to see the map in progress. Mangin flatly refused, and when Loss persisted, he wrote to the council to explain that the map was irrelevant to the harbor because it "is not the plan of the City such as it is, but such as it is to be."[10] If the councillors had been attentive, this certainly should have raised some concerns about their surveyor's intentions—concerns that were fully validated when they were handed his completed chart. They had just spent $3,000 in public funds for the purpose of producing a detailed map of their city, and instead they were presented with a map of a city that existed only in Mangin's imagination. Yet the councillors were, on the whole, surprisingly nonchalant about the strange turn their commissioned survey had taken. Perhaps they thought that the part of the map that did describe the city could still serve their purposes, perhaps they were swayed by Mangin's eloquence, or perhaps they feared provoking the displeasure of his patron, Hamilton. Whatever the case, they did not complain, paid Mangin the remainder of his fee, and referred the map for engraving by master craftsman Peter Maverick so that it could be reproduced and sold to subscribers.[11]

Mangin's map was a remarkable document for more than one reason. The chaotic streets and blocks of the city were transformed

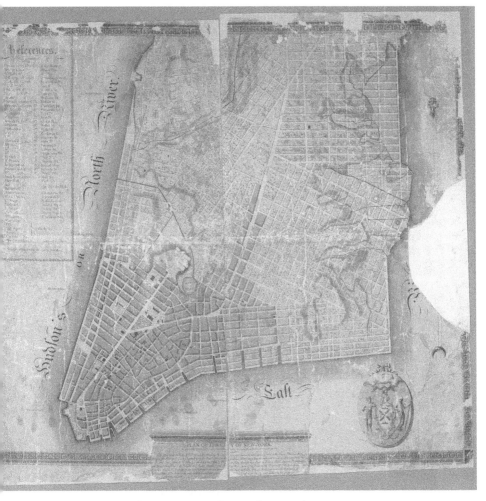

The Mangin-Goerck map of New York City, 1803.

into regular squares and rectangles, and the narrow southern tip of Manhattan was considerably broadened and expanded, making room for several new streets on the waterfront that did not—and could not—exist at the time. The waterfront itself, naturally ragged and irregular, was smoothed into straight lines that changed direction only in specific locations and at precise angles. How this expansion and smoothing of southern Manhattan would be accomplished, Mangin did not say. Most likely he imagined a massive public works project that would expand the area of the island with landfill. It would take many years, but at least some of the streets imagined by Mangin in southern Manhattan, including famous South Street, were eventually laid out on land reclaimed from the Hudson and the East River.

In the areas where New York was already built, Mangin's map expanded and improved, but for the most part it presented the city as it was—if more elegant and orderly than it appeared on the ground. But in the areas beyond the settled city, extending as much as two miles to the north and reaching into the area of Greenwich Village, he was much bolder. Here Mangin simply invented new avenues, streets, and entire neighborhoods, covering an area roughly twice the size of 1799 New York. The future city, as Mangin imagined it, would be neither the street jumble of old New York nor the massive uniform grid of the future. Instead it would be a patchwork of local grids, each regular and consistent in itself but oriented in a different direction from neighboring grids and with somewhat different dimensions. The boundaries between the adjoining grids became broad avenues that intersected one another in acute angles, creating natural plazas and triangular open spaces to serve as city parks.

Looking at Mangin's map, we can only imagine what it would have been like to walk the streets of this New York that never was. From the cozy gridded neighborhoods, one would step out into the broad boulevards and walk by well-trimmed city parks toward the angular plazas scattered about the city. Looking up, one would catch sight of a grand public building or great church that would no doubt be located where the boulevards meet, and viewable from many blocks away. With its broad avenues, parks, and plazas graced with civic monuments, Mangin's New York would have possessed the charm of a great European city such as Paris, or perhaps the most European of American cities—Washington, DC.

This should come as no surprise, for Mangin, like L'Enfant, had learned his craft in France, and it was only natural that he would attempt to transplant the traditions of European city planning onto American soil. Indeed, Mangin's attachment to French royal architecture was also fully in evidence in his proposal for New York's city hall, which resembles nothing so much as the Hôtel de Ville of a great city in France. Even after authorities forced him to tone down his design, it would still have been perfectly at home on a Parisian boulevard. It is more than likely that if Mangin had had his way in New York, similar edifices in similar style would have dotted the city's plazas, with grand avenues leading up to them.

Needless to say, it never happened, and it is hard, in retrospect, to imagine how exactly Mangin imagined his plan could be put into effect. He was never asked to produce a plan for a future New York and never requested permission to work on one; he simply presented it to the city councillors as if it were the fulfillment of his commission. His influence, and more likely that of Hamilton, was great enough in 1799 for the councillors to go along with the charade. But in 1803, when the engraved maps were printed and sent to subscribers, circumstances had changed dramatically.

The election of Thomas Jefferson as president in 1800 struck a stunning blow to the Federalists who had dominated American politics in the 1790s, and particularly to their leader, Hamilton. Over the following years, their positions of power were slowly but inexorably taken over by Jefferson's Republican Democrats, who espoused universal (white) male suffrage and despised the Federalists' aristocratic leanings. In this shifting environment, Mangin's design, inspired as it was by the royal capitals of Europe, would have seemed out of step with the new political winds. Add to this the fact that the influence of Hamilton was on the wane, and one gets a sense why the councillors no longer felt obliged to turn a blind eye to Mangin's transgressions. For there was no denying that what he had produced was not what he had been paid to do.

"The Map of the City lately printed," reported a select committee appointed by the council, "contains many inaccuracies and designates streets which have not been agreed to by the Corporation and which it would be improper to adopt, and which might tend to lead the proprietors of Land adjacent to such streets so laid down into error." It was a devastating indictment of what had been intended to serve as the definitive and official map of New York, but the councillors did not dispute it. Instead, they asked that all the maps that had been delivered to subscribers be recalled and that all the fees already paid be refunded. They also decreed that those who wished nonetheless to retain the map would have it pasted with a warning label, specifying that Mangin's street plan was entirely nonbinding and that the future development of the city would be determined by the council alone.[12]

For Mangin's imagined New York, this was the end of the road. For all its originality and Old World charm, it remained a city on paper only. As New York expanded rapidly in subsequent decades, while following a very different street plan, it became a ghostly reminder of a city that never was and a path not taken. And yet, despite never being more than a paper chart, Mangin's plan was a turning point for the city of New York. It represented the first time in the history of the city that anyone had proposed a concrete and detailed plan for its future development. Typically for a city whose leadership was distinguished mostly by its passivity in the face of assertive private interests, the plan did not originate in the Common Council but rather in the personal initiative of a renegade surveyor with a high opinion of his own talents.

The Common Council rejected Mangin's plan for the future, but it could not go back to the past. Ever since its early beginnings as New Amsterdam, the city's authorities had done their best to avoid planning ahead for New York's inevitable growth, as any such plan would have required taking on the vested interests of landowners. And if we consider that the city councillors and the chief landowners were often the very same people, it becomes clear why New York was left to expand chaotically, without plan or vision, for so many decades. But once Mangin had presented his unsolicited proposal, even the most reluctant officials could not put the genie back in the bottle. In explaining why they turned down the plan before them, they were forced to make the case that it was deficient and that a better one was possible. As they explained in the warning label they ordered attached to Mangin's maps, new streets were "to be considered subject to such future arrangements as the Corporation may deem best calculated to promote the health, introduce regularity, and conduce the convenience of the city."[13] Producing such a plan became the city's top priority.

The Commissioners

For several years the councillors cast about for someone to take on the design of future Manhattan. Initially, they charged a succession of their own street commissioners with surveying the island and proposing a street layout, but these proved unequal to the task. Dependent as they were on the city's power brokers, they were unwilling and unable to confront the entrenched interests of Manhattan's major landowners. Finally despairing of making headway through New York's weak and malleable Common Council, Mayor DeWitt Clinton turned to the state legislature in Albany, where longer terms of service and further remove from the city's fractious politics made prospects for action more promising. Clinton had additional reasons to believe that his appeal would be given full hearing at the state capital, for apart from his position as mayor of New York, he was also a state senator and one of the most powerful politicians in Albany. And so when in February 1807 he petitioned the legislature to pass a state law appointing a planning commission for New York City, he addressed his request to the most sympathetic recipient—himself.

Not surprisingly, the mayor's appeal was well received. On April 3, 1807, the last day of its session, the legislature passed "An Act relative to Improvements, touching on the laying out of Streets and Roads in the City of New York, and for other purposes." It was, needless to say, exactly the law that Clinton had asked for. To break through the gridlock of city politics, the act appointed three "Commissioners of streets and roads" with near-dictatorial powers over Manhattan's inhabitants. "The said Commissioners," the act read, "shall have and possess exclusive power to lay out streets, roads, and public squares of such width, extent, and direction, as to them shall seem most conducive to public good, and to shut up, or direct to be shut up, any streets or parts thereof which have been heretofore laid out."[14] These powers, it should be noted, did not apply to the old city at the southern end of Manhattan but only to the area beyond a jagged and carefully delineated line that today can be traced as the irregular southern boundary of the city's Great Grid. North of that line, however, over more than five-sixths of the length of the island, the commissioners' authority was absolute. Neither existing streets, local needs, neighborhood tradition, or—most troublingly—property rights, would stand in their way. In fact, the state act continued, it was lawful for the commissioners and "for all persons acting under their authority to enter, in the day time, into and upon any lands, tenements, or hereditaments which they shall deem necessary to be surveyed, used or converted for the laying out, opening, and forming of any street or road."[15]

The only limitation on these vast powers was one of time: "The powers and duties hereby given to the said Commissioners, shall be exercised and discharged by the said Commissioners within four years of the passing of this act, and not after."[16] On April 3, 1811, the commissioners' powers would expire, and they would be required to present their plan to the assembly.

It is hard to imagine that, in our present day, such unlimited powers assigned by the state to the commissioners would have survived a constitutional challenge. Not only did the commissioners have the right to close existing streets on private land and lay down new ones as they saw fit but anyone acting for the commissioners could enter any private land in Manhattan and legally set up their surveying instruments in a pasture, in a front yard, or even in someone's kitchen if they found that it served their needs. More than a street-planning bill, the law is a blatant power grab by the state, using a simple majority in the legislature to override the rights of Manhattan's citizens to their property, their privacy, and their person. But in 1807 legal precedents were few, and the power of the judiciary to override the actions of a state legislature was still very much in question. And so the law was passed with little opposition and immediately put into effect.

The New York legislature not only defined the commissioners' powers and responsibilities but also named them, once again following DeWitt Clinton's advice. Significantly, none of the three was a resident of Manhattan or a member of the city's political elite, which had scuttled the previous attempts at city planning. This was surely as Mayor Clinton had intended. The least well-known of the three, in his own time as in ours, was also the least influential and subject to persistent complaints from his fellow commissioners for failing to attend their meetings.[17] He was John Rutherfurd, a wealthy lawyer, landowner, and former US senator for New Jersey. Although he had been born in the New York and had lived there for a few years as an adult, he had long since settled in his family estate in Sussex County, New Jersey.

Far more prominent was Simeon De Witt, surveyor general for the State of New York. De Witt had served on Washington's staff during the Revolutionary War and had been named "Geographer to the Main Army" and later "Geographer to the United States of America," along-side Thomas Hutchins. De Witt remained close to Washington, who recommended him to Jefferson in 1784 as a "Modest, sensible, sober, and deserving young Man, Esteemed a very good Mathematician, and well worthy encouragement."[18] Whether Jefferson was impressed we don't know, for he soon left for France without being able to do any-thing for Washington's protégé. When later that year Congress was gearing up to launch what became the survey of the Seven Ranges, it is likely that De Witt was Washington's preferred candidate to lead it. Instead, De Witt chose to resign his position as Geographer to the United States and accept an appointment as surveyor general to his home state of New York, leaving the survey of the Ohio Valley to Hutchins.

As surveyor general for New York, De Witt was not idle. In 1791, he surveyed the lands of the new town of Ithaca, which he named, and laid down the streets in a classical grid pattern. Three years later, as the national capital was taking shape far to the south, he laid down a new street plan for New York's state capital, Albany. Impressed, perhaps, with L'Enfant's design, though on a far smaller scale, he combined a traditional grid with grand boulevards converging on the state house, in the manner of European capitals.

De Witt's greatest accomplishment, however, was his map of the

State of New York, which was completed in 1802 and remained the standard map of the state for decades to come. Inspired by the national survey taking place beyond the Appalachians, he carved western New York into regular rectangles that ignored natural features and Indian habitation. Like Hutchins and his successors, De Witt did his best to present the land as a blank slate, inviting new settlers to put their mark on the supposedly virgin land. New York, however, with its relatively dense habitation and long history of interactions between Europeans and native peoples, proved less amenable to such treatment than the Great Plains. In the end De Witt produced a patchwork of grids, only some of which were oriented to the points of the compass and all of which were forced to accommodate the very non-rectilinear shape of New York's counties. Even so, De Witt's map set a new standard of precision and detail in state maps and established his reputation as one of the leading surveyors in America. When in 1807 his relative, Mayor DeWitt Clinton, was looking for someone to survey and lay out the streets of New York City, Simeon De Witt was an obvious choice.

The Rake with the Wooden Leg

The most influential as well as the most colorful member of the commission, the one whose views carried the most weight and who wrote the commission's final report in 1811, was Gouverneur Morris (1752–1816). Born on his wealthy family's estate of Morrisiana, in what is now a district in the southern Bronx, he distinguished himself early for his sharp mind and personal charisma. The first of these would see him acquire a master's degree from King's College (future Columbia University) at the age of nineteen and gain admission to the bar in New York at twenty-four. The second would earn him a lifelong reputation as a rake whose affairs with women, both married and unmarried, spanned continents. In 1780, Morris lost his left leg after being thrown off his carriage and catching his ankle in the spokes of one of its wheels. This was likely a simple accident, but a rumor soon spread that he had suffered his injury while trying to evade an angry cuckold.

Morris himself did nothing to discourage such talk, even hinting that the peg leg he wore for the rest of his life was an asset in his amorous adventures. "Gouverneur's leg has been a tax on my Heart," wrote his friend and fellow New Yorker John Jay to Robert Morris some months after the accident. "I am almost tempted to wish he had lost *something* else."[19]

In 1775, as the rupture with England was reaching its crisis, Morris broke with many in his family, including his mother, and sided decisively with the patriot cause. Elected to the New York State Assembly, he was forced to flee first Manhattan and then his family estate, ending up with his fellow delegates in the town of Kingston in the Hudson Valley, where he took part in drafting the state constitution. After the Battle of Saratoga, as military operations in New England drew to a close, Morris was appointed a delegate to the Continental Congress, which he joined in its exile in York in January 1778. He was immediately sent to report on the condition of the Continental Army and was so appalled by the conditions he witnessed in its winter encampment at Valley Forge that he became the chief advocate for the army in Congress and a key ally of its commander, Washington. Morris spent only two years in Congress before being defeated in his bid for reelection, and when he lost his leg later that year, his public career seemed to be at an end. But in 1781 he accepted the invitation of Robert Morris, newly appointed superintendent of finance for the United States(though no relation), to serve as his assistant.

The elder Morris had arrived from England decades before as a penniless young man, but by this time he was famous as a wily trader and speculator who had no equal this side of the Atlantic. As one of the wealthiest men in America, he was charged with keeping the rickety finances of the new nation afloat during its years of struggle, and he did so in no small part by paying for the war effort out of his own pocket. Working together to secure the republic's credit in the world, the two Morrises proved to be a powerful team. Both believed that the future prosperity of America depended on the success of its merchants and bankers, and both believed that a powerful central government was required to safeguard their interests. To Jefferson, who disliked both men, they were the embodiment of "that Speculating phalanx" who would turn his "empire for liberty" into an Eden for greedy financiers. Even when Jefferson agreed with the Morrises on the need for reform, he could never bring himself to cooperate with them to accomplish it. When young Gouverneur proposed a currency reform that would substitute the plethora of coins being used in the states with a US currency based on the Spanish dollar and its division into eighths, Jefferson immediately countered with his proposal for a decimal dollar. On this occasion, he won the day.

In 1785, Morris retired to Philadelphia, but his reprieve as a private citizen turned out to be brief. The very next year, he was appointed by the Pennsylvania General Assembly to serve as one of the state's delegates to the Constitutional Convention. Morris claimed to have been surprised by his appointment, but it soon turned out that the convention setting was an ideal platform for a man of his character and skills to shine. A natural provocateur, with an easy wit and a talent for self-promotion, Morris rose repeatedly to speak in defense of a strong national government and to denounce all efforts to vest powers in the states. "What if all the charters and constitutions of the states were thrown into the fire, and all their demagogues into the ocean? What would it be to the happiness of America?"[20]

Perfectly in line with his opposition to the diffusion of power to the states was his concern over what he called an "excess of democracy." If the common people were allowed full voting rights, he warned, "the rich will take advantage of their passions and make these the instruments of oppressing them."[21] Yet when it came to slavery, no delegate at the convention took on the issue as openly as Morris or denounced its defenders more passionately. On the delicate issue of counting slaves in the drawing of congressional districts, he fiercely denounced the rule by which "the inhabitant of Georgia and South Carolina," who cruelly enslaved his fellow men, "shall have more votes in a government instituted for the protection of the rights of mankind than the citizen of Pennsylvania or New Jersey who views with a laudable horror so nefarious a practice."[22]

When all was said and done, Morris rose to speak more often

than anyone else at the convention, leaving a powerful impression on his fellow delegates. This, to be sure, does not mean that he was the most influential in shaping the final outcome of the deliberations. James Madison and Alexander Hamilton, who spoke less often but spent their time forging alliances and brokering compromises, had far greater say in determining the contents of the Constitution than did Morris. But when it came time to combine the convention's disparate resolutions into a single document, the delegates were impressed enough with his eloquence to select him to a five-man committee charged with drafting the final text. Morris did not disappoint and is credited with streamlining and refining the awkward legalistic phrasing of the original resolutions into a coherent text couched in elegant prose.

Morris's most enduring contribution to the convention, as well as to American history, came at the tail end of the committee's work, when he was charged with composing the preamble to the Constitution. In a few simple sentences, he did more than summarize the purpose and meaning of the new Constitution. He redefined the Union, casting the residents of the various states as a free and sovereign people, united in a common destiny.

"We the People of the United States, in Order to form a more perfect Union, establish Justice, insure domestic Tranquility, provide for the common defence, promote the general Welfare, and secure the Blessings of Liberty to ourselves and our Posterity, do ordain and establish this Constitution for the United States of America."

The words have resonated through the centuries and are familiar to American schoolchildren to this day. And they earned Gouverneur Morris—rake, womanizer, and provocateur though he was—the unofficial but enduring title of Founding Father of the United States.

By the time Morris was appointed as one of the three commissioners charged with planning the future of New York City, his triumphs in the Constitutional Convention were two decades in the past. In the intervening years, he had traveled to Paris, first on private business and later (from 1792 to 1794) to serve as Washington's minister plenipotentiary to France. Unlike Jefferson, who had witnessed the early stirrings of the revolution and then cheered it on from afar, Morris experienced it in full swing and was troubled by what he saw. Always suspicious of "excessive democracy" and fearful of mob rule, he viewed the increasing radicalization of the revolution with trepidation. He did his best to help the beleaguered royals, as well as his many aristocratic friends, in their time of need but was helpless to stop the rising tide of violence and terror. After being removed from his official position at the request of the Jacobin government, he traveled extensively in Europe before landing back in the United States at the end of 1798. Drawn once more into the political fray, he became a Federalist and served for two years as a US senator from New York. Defeated in his quest for reelection, he retired permanently from the national political scene.

In 1807, Morris was perhaps the best-known and most influential New Yorker of his day, though he held no official post. A member of one of New York's oldest and most prominent families, he was a landholder, a successful financier, and one of the wealthiest men in the state. As a former minister to France and US senator, he knew everyone in the nation's political elite and counted many of them as intimate friends. But it was his role as one of the chief architects of the Constitution that enhanced Morris's stature beyond all measure in the eyes of his contemporaries. The man who coined the phrase "We the people" could not but be considered one of the giants of the age.

It was likely Morris's towering public stature that persuaded DeWitt Clinton and his colleagues in the state assembly to appoint him to the New York Street Commission. Redrawing the chaotic map of New York City and planning its future expansion, after all, was a grandiose task, whose chances of success were murky at best. Whatever plan the commissioners eventually settled on was certain to arouse fierce opposition among the wealthiest and most powerful men in the city. If the commissioners' vision for Manhattan was to prove more than just another paper plan, if there was any hope of it being actually implemented on the ground, it would require backing from a man whose authority had few equals. And this indeed was what Morris brought to the commission. John Rutherfurd was respectable and wealthy but rather obscure; Simeon De Witt brought his professional expertise and a distinguished track record as state surveyor and mapmaker of New York; but only Morris brought the stature and gravitas of a Founding Father of the United States. Without him, no plan would have much hope of making the leap from the drawing board to the city streets.

The Surveyor

On April 4, 1807, with no fanfare whatsoever, the New York State Assembly approved the "Act relative to Improvements, Touching On the Laying Out of Streets and Roads in the City of New York, and for Other Purposes."[23] It was an unassuming title for an act that would forever reshape the greatest metropolis in North America. The title gave no hint of the extraordinary powers invested in the three commissioners named in the act. It did not mention that they were accorded the right to override property rights throughout the island, enter private lands to conduct their business with or without the owner's permission, and lay down streets and roads however they saw fit. And it is likely that the authors of the act preferred it that way, so as not to draw attention to its explosive contents. Nevertheless, as early as January 1807, three months before he was officially appointed commissioner, Gouverneur Morris already understood the implications of the work before him: "It seems as if the whole island of New York were soon to become a village or a Town," he wrote. "In less than 20 years . . . it will be divided in small Lots as far up as what are called Haerlem Heights where stood Fort Washington."[24] We do not know whether Morris knew of his coming appointment when he wrote this, but over the next four years he would be the one most responsible for turning it into reality.

Our records of the work of the commissioners who would determine the shape of New York City are surprisingly sparse. Much to the distress of future historians, Morris and his colleagues did not keep minutes of their meetings, and we are left to piece together what they did and when from Morris's entries in his personal diary, the field

notes of the surveyors employed by the commissioners, and anec-
dotes related decades later by descendants and acquaintances of those
involved.[25] There is no question, however, that the work got off to a
slow and rocky start.

Shortly after their appointment, the commissioners hired Charles
Loss as their chief surveyor and charged him with preparing an accurate map of the island of Manhattan. We have already met Loss, who
had clashed with Mangin some years before, and with the Frenchman
and his plans out of favor, Loss had emerged as one of New York's
leading surveyors. Unfortunately for Loss, and for the commissioners who had hired him, Loss's skills as a surveyor did not match his
impressive talents for local politicking. By the fall of 1807, Morris was
complaining to De Witt about Loss's unreliable work habits. Things
had gotten so bad that he had taken to playing a cat-and-mouse game
with the surveyor, dropping by unannounced at his field office. "My
arrival being uncertain," he wrote, "makes it inconvenient to Mr. Loss
to absent himself."[26]

When it came to the work the surveyor did perform, things were
no better. Loss, according to Morris, was relying on old maps, without verifying that their north–south orientation was true and aligned
with the meridian. When Loss, at Morris's insistence, compared the
different maps, he predictably discovered that their meridians differed significantly from each other and were therefore unreliable. The
commissioner then instructed Loss to do what he should have done
in the first place—measure the variation of the compass, use it to
determine the true direction of the city's streets, and map them accordingly. All the maps that Loss had produced to that point now had
to be discarded. "We cannot indeed remedy What is past," a clearly
frustrated Morris wrote to De Witt, "but mischief may be presented
in the future. Will you my dear Sir pardon me for suggesting that
an official Representation from the Surveyor General might not be
improper."[27]

Whether De Witt visited Loss and tried to correct his surveying
habits we may never know, but time was clearly running out for the
chief surveyor. Arrangements were made that would see Loss's daily

wages as surveyor converted to a fixed sum, to be paid when his map of the island was complete. When in May 1808 he submitted his finished map, he was duly paid, and his services to the commissioners were at an end. Considering how little faith Morris had in the quality of the surveyor's work and how doubtful it was that the map he had produced would actually be used, the terms of the settlement were surprisingly generous. Once again it seems that Loss's penchant for falling on his feet benefited him more than his surveying skills.

More than a year after the commissioners' appointment, the project of producing a street plan for the future growth of New York City had barely gotten off the ground. Months had been wasted on producing a map of Manhattan as it currently was, and it was of dubious usability. Nothing at all had been done to advance the commission's ultimate goal—marking out the streets of the city as it would one day be. To get the project moving again, Morris and his colleagues most urgently needed a new chief surveyor, one who would correct Loss's errors and mark out the location of the future streets. They needed a man who was not only a competent surveyor but also a reliable one, and one who would take direction from the commissioners and work closely with them. Learning from past experience, they steered clear of established city surveyors, men who might have their own agendas and their own political connections, not to mention other competing commitments. Instead they brought in from Albany a young surveyor of limited experience, but one they knew well and whose skills they trusted. Most importantly, he was a man for whom the appointment as chief surveyor for the commission would be a giant career leap forward and who owed his position and prominence entirely to them. His name was John Randel Jr., and he was only twenty years old.

John Randel was Simeon De Witt's man. Born in Albany in 1787, John likely knew the state surveyor from a young age. The Randels were apparently well acquainted with the De Witt family and valued their connection so much that they named one of their daughters Jane DeWitt. When John distinguished himself early for his mathematical talents, Simeon De Witt took him under his wing and introduced him to the science and art of surveying. By the time Randel was a teenager, he was working regularly in De Witt's office, and was soon trusted to complete independent surveying assignments. He made maps of parts of the Adirondack Mountains, a section of the Oneida Indian Reservation, and the Great Western Turnpike from Albany to Cooperstown. By 1807, under De Witt's supervision, he was carving up portions of central New York into a regular rectilinear grid. And so, when the commissioners needed a surveyor who was competent, reliable, and unquestionably loyal, De Witt knew the man for the job. By July 1808, young Randel and his surveying instruments were in Manhattan, working day and night as the commission's newly minted chief surveyor.

What exactly Randel was surveying in these early days is once again not easy to determine. His forty-five field books, preserved today at the New York Historical Society, cover his work for the decade and a half between 1808 and 1823 and vary enormously in the detail they provide. Unfortunately, they provide only a rough idea of his activities in the early years after his appointment as the commissioners' chief surveyor. Part of his work was probably done to correct the shortcomings of Loss's map of the island as it currently was. He mapped out existing roads, such as the Eastern Post Road, which ran along the eastern side of Manhattan, as well as Greenwich Lane, Art Street, and others. He carefully noted the lay of the land and its nature, marking hills, rocks, ponds, streams, woods, marshes, and so on. This detailed mapping of the island as it was should, no doubt, have been completed by the time Randel arrived on the scene. The fact that he devoted so much time to it is a measure of how badly his predecessor had bungled the job.[28]

Beyond mapping the topography and roads of Manhattan, Randel also focused his measurements on the "Common Lands," surveyed by Casimir Goerck more than a decade before. Goerck, it will be recalled, divided the lands into a checkerboard of rectangular five-acre plots, with each plot enclosed by broad avenues paralleling the island's east–west axis and narrower east–west streets that intersect them at right angles. We do not know when exactly the commissioners made the fateful decision to extend the Goerck grid to the entire island and thereby create the most famous urban landscape in the world. But Randel was undoubtedly following the instructions of Morris and De Witt when he slowly and systematically extended the outlines of the Common Lands to the island's open countryside and the shores of the East and Hudson Rivers. Even if the commissioners had not yet decided on their final plan, Randel's surveying notes of 1808–10 suggest that they were seriously entertaining the possibility of a gridded city. The notes are, in a way, a record of the birth pangs of the Manhattan grid.[29]

In later years, Randel would develop a reputation for superb precision in his surveys, above and beyond what was expected in his day. He even went so far as to design and build his own surveying instruments when he judged that the ones available for purchase were not up to his exacting standards.[30] But between 1808 and 1811, when he was but a young surveyor working under the commissioners' directions, his methods were simple enough and closely resembled the practices of Jared Mansfield's men, who were laying out the Great Western Grid.

To lay out a straight line in a given direction, Randel would set a theodolite at the starting point and aim it in the desired direction. A theodolite includes a sight, which is usually a telescope (though that was not always the case in the early nineteenth century) set on a compass that allows one to determine the general direction one is aiming. Looking through the theodolite in the desired direction, Randel would select a distant landmark and send members of his crew there to plant a visible marker. Teams of flag men would communicate between the starting point and the landmark to make sure that the marker was installed at precisely the right location, and axmen would clear the path between the two points. Then the chainmen, wielding sixty-six-foot-long Gunther chains, would proceed in their characteristic caterpillar movement: stretching out the chain from the starting point, putting down a marker at the end, and then repeating the process from that point, time after time, until the end marker was reached. Throughout the process, Randel or an assistant would keep the team in the theodolite's sights to make sure they were proceeding in a straight line. When the process was complete, a precisely measured straight line of known length would connect the beginning point and the marker.

Determining the exact true direction of the line, however, was somewhat more challenging. The simplest means was by using a compass, but a compass never points to the true geographic north: it points to the magnetic north, which varies substantially over time. In addition, compass readings are influenced by local geological formations and vary significantly from one location to another, even within a small area such as the island of Manhattan. Well-known though it is, magnetic variation is not easy to overcome, and correcting for it is intensely time-consuming. It requires using astronomical methods to determine the true north and then measuring the compass's deviation from it at different locations. That Loss had ignored these basic facts led to the premature end of his tenure as chief surveyor. Randel, not wishing to follow in his predecessor's footsteps, paid close attention to the issue as he laid out the Manhattan grid. After all, if the streets of New York were to line up parallel to one another and at right angles to the avenues, then they would all need to point in precisely—not approximately—the right direction.[31]

Not only was the survey of Manhattan technologically challenging but it also required intense physical labor. Randel and his men traipsed up hills and down valleys, through marshes, over rock outcrops, and across creeks and lakes. When traversing the bush, they would need to clear paths for the chainmen. When crossing a wood, dozens of trees would be felled to provide clear lines of sight. And so it went, day by day and year after year, up and down the island of Manhattan. It is probably a good thing that Randel was a young man when he started out on his survey, and it may be that his youth actually commended him to the commissioners. An older man might not have had the stamina for such an undertaking.

And yet it was not the physical challenges of the survey that did most to slow down Randel and his men. It was, rather, the human challenges, or more precisely the intense hostility of the landowners whose properties they traversed. One can easily imagine the horror of a Manhattan farmer when they spied Randel's team approaching in the distance. Armed with their authority "to enter, in the day time, into and upon any lands, tenements, or hereditaments which they shall deem necessary to be surveyed," the surveyors would invade fields, meadows, and orchards, paying no heed to the protestations of the owners or tenants. They laid down their instruments wherever they wished and immediately set upon clearing the brush, chopping down trees, and breaking through hedges and fences wherever they saw fit. Randel and his men were out to establish straight lines, and whatever stood in their path would have to give way.

Little wonder that, faced with such high-handed violation of their rights, the landowners fought back. "At the approach of engineers with their measuring instruments, maps, and chain-bearers," recounted Martha Lamb in her 1877 *History of the City of New York*, "dogs were brought into service, and whole families sometimes united in driving them out of their lots, as if they were common vagrants."[32] On one notorious (and possibly apocryphal) occasion, the surveyors were set to draw a line right through the kitchen of an "estimable old woman" who sold vegetables for a living. Rather than submit to their intrusion, the lady pelted Randel and his men with a shower of cabbages and artichokes, forcing a hasty retreat "in the exact reverse of good order."[33] Another popular tactic by landowners was to steal or damage the markers Randel's men left in the ground to track their progress or pinpoint future intersections.[34] But when dogs, artichokes, and even vandalism proved insufficient to stop the surveyors' advance, the people of Manhattan turned to their ultimate line of defense: they called on the law and turned it loose on the commissioners' men. "I was arrested by the Sherriff, on numerous suits instituted against me," Randel recalled many years later, "for trespass and damage committed by my workmen, in passing over grounds, cutting off branches of trees, &c., to make surveys under instructions from the Commissioners."[35]

As Randel tells it, his men did nothing more than cut off branches where necessary. The landowners, for their part, saw things differently. In a suit brought by an upstanding citizen named John Mills, he details the wholesale destruction wrought on the Mills farm on

August 26, 1808, when it was visited by Randel and his men. The surveyors, the suit charges, with "force and arms to wit with hands, feet, swords, staves, sticks, stones, knives and axes" cut down, trampled, and destroyed "five hundred Ash Trees, Five hundred yew Trees, Five hundred Elm Trees, five hundred Apricot Trees, five hundred Peach Trees," and on and on. Mills's account of the damage he suffered seems wildly exaggerated, as does the $5,000 he demanded in compensation (he was ultimately awarded $109 and 63 cents). But the fury at the surveyors, which comes through in his suit and which he shared with many of his neighbors, was real enough.[36]

There was more, however, to the landowners' hostility than concern about physical damage to their property. For nearly a century and a half, ever since New Amsterdam became New York, landowners had effectively ruled the chaotic city. The Dongan Charter of 1686 had in theory given New York's Common Council authority to administer the entire island of Manhattan, but the reality on the ground was very different. In practice, property holders were left to do much as they pleased and lay down streets and roads as they saw fit, with no regard to the broader good of the city's inhabitants. Given that the city councillors themselves were members of this class and that they relied on the goodwill of their neighbors to retain their seats in annual elections, this situation was unlikely to change. And that was just how New York landowners liked it.

But when John Mills and his neighbors spied the commissioners' teams of surveyors advancing on their lands, they sensed that the days of their ascendancy were numbered. To them, Randel and his men were more than unwelcome intruders who trampled their field and felled a few trees; they were, rather, the vanguard of a new regime that would forever change the way things were done in New York. Their mandate was to establish a unified system of streets and roads, which would override and ultimately replace the local arrangements of private property holders. Just as ominously, they were armed with an act of the state assembly and were thereby immune to the pressures of city politics. The surveyors, in other words, were the harbingers of a political and administrative revolution that would forever end the political dominance of New York's old families. The old elite saw them as such and determined to fight—by dogs, by hails of vegetables, and, most effectively, by the local law courts.[37]

By the summer of 1808, opposition had slowed the surveyors' work so much that De Witt warned the Common Council that if they were not protected from "vexatious interruptions," the commission would be unable to complete its duties on time. The city councillors, as usual, proved unable or unwilling to support Randel against his tormentors. The job of resolving the impasse was left once again to the state legislature, which in March 1809 passed a new resolution authorizing Randel to conduct the survey as he saw fit. The new act reaffirmed the surveyors' right to enter any property on the island during the day, this time stating explicitly that they had the authority to cut down trees and inflict "other damages" if no other way could be found to conduct their work. Most significantly, if sued by property owners, Randel and his men "could give this act" as evidence in their defense. In other words, regardless of the court in which they were sued, the surveyors were now protected by state law.[38]

After this, things seem to have gone more smoothly for Randel and his men, and by the summer of 1810, he had settled on the outlines of the future city, including the locations of the avenues and streets. We know this because Randel spent September of that year away from Manhattan, surveying the Post Road between Albany and New York and producing a detailed map, twenty-three feet long. The southern portion of the map shows the road entering New York City, leading up to city hall. In what might be the earliest draft of future Manhattan, the road winds its way through a perfectly gridded city, unmistakably the one depicted in the official Commissioners' Plan some months later.[39]

By early in the winter of 1810, even though much work remained to be done on the survey and mapping of Manhattan, the commissioners felt confident enough to report on their plan to the Common Council. "I am directed to inform you," Morris wrote to the council on November 29, "that the Commissioners for laying out the Manhattan Island have completed their work so far as depends on them; but much is yet to be done on the ground." Morris went on to explain that the general outline of the plan, which he considered the commissioners' responsibility, was close to completion: "It will be practicable," he wrote, "to make within the time fixed by the Statute a report complying substantially, if not literally with the law, shewing all the streets to be laid out." But whereas the paper outline was close to completion, the work on the ground had barely begun. "There are," Morris noted, "measured and to be measured . . . upwards of five hundred and fifty thousand feet, that is to say upwards of one hundred miles." An additional challenge was placing 3,500 markers, or "monuments," as he called them, at all projected intersections so that the streets and roads could be laid out in the future. By the time the commissioners submitted their report in March 1811, only 30 monuments had been placed.[40]

All this work, needless to say, would still take years to complete, long after the commissioners had presented their plan to the state legislature. There was only one man, according to Morris, who was up to the task: "I am directed to state," he wrote the council, "that only one surveyor can be employed in what remains . . . where the difference of an inch may afterwards be a source of contention." That, of course, was John Randel. At Morris's instigation, the council did indeed hire Randel, and he spent the next decade traipsing up and down Manhattan Island, completing the job he had begun under the commissioners' direction.

He was, without question, the man for the job. By the time he was done, in 1821, he had surveyed over seven hundred thousand feet of streets and avenues and placed over 1,600 "monuments" at projected intersections (significantly fewer than Morris's estimate of 3,500). He had produced a map of the Commissioners' Plan for Manhattan in three copies for the state legislature in 1811 and another improved version of the map in 1814 (which was ultimately published in 1821). Most impressively, between 1818 and 1821 he produced ninety-two separate maps, each thirty-two by twenty inches and showing different parts of Manhattan on a scale of one inch to one hundred feet. When taken together, they form a map, about fifty feet long, of the entire island in exquisite detail. In 1808, on the recommendation of Simeon De Witt, the commissioners hired twenty-year-old John Randel to survey the island of Manhattan for them. By the time he had completed his survey thirteen years later, the inexperienced youth of 1808 had become one of the top surveyors in the United States.[41]

The Ultimate Grid

On March 22, 1811, just thirteen days before the expiration of their mandate, the commissioners officially filed their report. As required by the 1807 act that established the commission, it included three copies of a street plan for Manhattan, all prepared by Randel. One copy went to the New York secretary of state, another to the clerk of the city and county of New York, and another to the Common Council of New York City. All three maps survive today in the custody of state and municipal archives and the New York Public Library, a sure sign of their enduring importance.

The maps are huge. Each one is nearly nine feet long, and when looked at closely, it includes many of the natural features of New York in 1811. The rugged landscape of the "Island of Hills" is all there, comparatively flat in the southeast of the island but increasingly uneven and hilly to the west and north, leading to jugged rocky Harlem Heights. Streams and creeks are clearly marked, pouring into both the Hudson and East Rivers, from Kip's Creek in the south to the more imposing Harlem Creek farther north. Most of the shoreline of the Harlem River, on the northeast side of the island, is marked as a bog called the Haerlem Marsh.

Human features appear on the map as well. Bloomingdale Road, which will one day become Broadway, is there, running up the west side, as is the Eastern Post Road, opposite it on the eastern side, and the Kings Bridge Road, winding its way up between them to the bridge at the northern tip of the island. Various private initiatives, such as Norton's Road and Steuben Street, are there too. The names of property owners are scattered throughout the map: Arden, Nichols, Taylor, Turnbull, repeatedly Stuyvesant—and tellingly, Morris. Numerous existing public buildings are marked as well: Mangin's sparkling

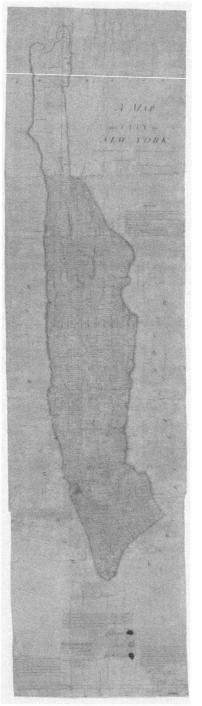

The Commissioners' Plan for Manhattan, 1811. From New York Public Library Digital Collections, Image ID 5125699.

new city hall (drawn but not named), an arsenal, a powder house, a bank, various churches throughout, and, suggestively, an establishment named Sailors' Snug Harbor.[42] It is all there.

And yet anyone looking at the map might be forgiven if they missed the natural and human features of the island entirely. For the map is dominated by one single construct that obscures all else: the great uniform and uncompromising grid that covers nearly all of Manhattan. The impression one gets from a brief look at the map is of an entire island covered by block after block after block of uniform rectangles, the same in every direction. Nothing interrupts the relentless monotony of the New York grid: the hills and valleys, creeks, meadows, and marshes are all covered by identical rectangles; the old roads, private properties, churches, and public buildings are lost beneath a uniform network of arrow-straight streets and avenues that intersect one another at regular intervals and right angles.

For all its apparent flatness, the commissioners' map of New York is, in fact two maps, distinguished by a clear timeline. One is the map of Manhattan as it was, based on Randel's careful survey. It shows the varied natural landscape of the island, as well as traces of its human history. The other is the map of Manhattan as it will be and is composed entirely of a uniform rectilinear grid. The two maps have nothing in common, except the outlines of the island itself, for the map of the future takes no account of the map of the past and present, refusing to accommodate hill, wood, or farm. One might expect that the uniform blocks would be shortened or modified to circumvent the rocky outcrops of Harlem Heights, but there is no sign of that. One might be concerned that the marshes along the East River would be an unsuitable terrain for building large structures or laying out streets, but there is no sign of that either. And there is certainly no sign of the grid conceding to the boundaries of any of the many private properties noted on the map. In fact, the map of future Manhattan is simply overlaid upon old Manhattan and replaces it, as if it were never there. The hills and farms of the present are drowned by the straight streets and avenues of the future, still visible beneath the uniform grid like ancient shipwrecks beneath a placid sea: quaint and charming in their way but entirely irrelevant.

With its two maps covering the same surface area but separated by time, the commissioners' map of New York City tells a story. It is a story of the island that had been and, in 1811, still was, an "Island of Hills," and creeks, and meadows, as well as villages, farms, fields, and orchards. And it is the story of what would happen to this island. For as the present turned to the future, this island would be erased and turned into something else entirely: a uniform featureless flatland, marked by a mathematical coordinate system of streets and avenues. "Strikers' Bay" on the west side will be merely the intersection of 11th Avenue and 96th Street, "Kip's Bay" on the East River shoreline will henceforth be known as 1st and 36th, while "Elgin Garden" will become an anonymous block circumscribed by 5th and 6th Avenues, and 29th and 30th Streets. What had been a rich and varied terrain, the map recounts, will soon be an abstract space in which every point is like any other, marked by the intersection of two numbered coordinates. A landscape that was natural will become mathematical.

Not all of Manhattan. The southern tip of the island, site of old New York, was outside the commissioners' mandate and therefore retained its confused and irregular street plan. But once the grid pattern was set in motion in the Commissioners' Plan, there was no stopping it. Its first signs appeared where tiny 1st Street, only three blocks long, still connects Houston Street (in 1811 "North Street") and the Bowery, just north of where the two intersect. From that point on the east side, the grid continues uninterrupted northward, but things are different out on the west side. There the grid does not make its appearance until tiny 12th Street, which exists today as a single block connecting Gansevoort Street with the Hudson River waterfront. The first river-to-river thoroughfare is 14th Street, and its cross-island pattern is then repeated in an additional 141 parallel streets right up to 155th Street. This was the northern boundary of the Commissioners' Plan, where they chose to end the grid, leaving the narrow northern quarter of the island unmarked.

The length of the Manhattan grid is defined by arrow-straight and parallel avenues that run most of the length of the island. They are aligned along the north–south axis of the island, which is, in fact, tilted twenty-nine degrees eastward of the true north and numbered from 1st to 12th: 1st Avenue runs along the East River shoreline, and 12th Avenue runs along the Hudson River shoreline, with the other ten avenues placed in sequence between them. On the east side, in places where the island is wider, four additional avenues are appended to the main grid, with Avenue A just east of 1st Avenue, and Avenues B, C, and D following eastward in sequence. Each avenue was to be 100 feet wide, but they were unevenly spaced. The avenues in the center of the island, from 3rd to 6th, were separated by 920 feet because they were extensions of the main roads that had been marked in Goerck's survey of the Common Lands two decades previously. But the gaps between the remaining avenues were smaller: 650 feet between 1st and 2nd, 610 between 2nd and 3rd, and 800 feet between the Avenues from 6th to 12th.

The east–west streets were in some ways more uniform. Numbered from 1 to 155, they were all laid out in a precise right angle to the avenues, and from 14th northward—unless interrupted by rectilinear squares—they traversed the island from river to river. All were parallel to one another and separated by 200 feet, and the vast majority were exactly 60 feet wide. The exceptions were fifteen designated east–west thoroughfares, which were to be, like the avenues, 100 feet wide. These were 14th, 23rd, 34th, 42nd, 57th, 72nd, 79th, 86th, 96th, 106th, 116th, 125th, 135th, 145th, and 155th Streets. Why the commissioners selected these particular streets has remained one of New York's unsolved mysteries.

There were, nonetheless, a few breaks in the unrelenting monotony of the grid. Most notably, a rectangular parade ground extended between 23rd and 33rd Streets and 3rd and 7th Avenues. This massive area, Morris explained in his remarks on the plan, was to be "set aside for military exercise, as also to assemble, in case of need, the force destined to defend the city."[43] The commissioners' concern for military defense is certainly understandable, as tensions with Great Britain (which were to lead to open war the following year) were on the rise, bringing back bitter memories of the British occupation three decades before. But the plan also made ample provision for peaceful economic activity, designating the elongated rectangle between 1st Avenue and the East River waterfront, from 7th to 10th Streets, as the city's "Market Place." As Morris explained in his remarks, concentrating sellers of "butcher's meat, poultry, fish, game, vegetables, and fruit" all in a single location, instead of spread out through the neighborhood, would help rationalize prices, increase quality control, and make commerce more efficient.

A few squares, each nestled between two avenues and several street blocks wide, are also sprinkled through the island, including one designated as the location of a future observatory, but that is all. Like the parade ground and the marketplace, they too are prisoners of the grid—rectangular in shape, with straight borders and right-angled corners, and occupying a precise number of city blocks. They do not so much disrupt the rectilinear world envisioned by the commissioners as reiterate it on a somewhat larger scale. Other than those limited exceptions, from 14th Street up to 155th there is no let up or respite from the relentless march of the conquering grid.

With their 1811 plan, the commissioners made the island of Manhattan into an abstract, featureless, and uniform space. Gone are the hills, valley, creeks, and marshes of the island; gone as well are the farms and villages that dotted the landscape. All of these would be ignored and, whenever necessary, removed. They have been replaced by a blank slate, empty and unresisting, in which every place is the same as any other, and each point is marked only by the rectilinear grid of avenues and streets. For that is, undoubtedly, what the street map of New York is: a mathematical construct based on a rectilinear coordinate system.

From a pure mathematical standpoint, the Manhattan grid is not perfect. Whereas the streets are more or less regularly spaced, like Cartesian coordinate points, the same cannot be said of the avenues, which are farther apart than the streets and of varying distances from one another. As a result, the grid's basic units, or blocks, are not square but rectangular, and while their width is constant, their length varies somewhat. Nevertheless, the key feature of an abstract mathematical plain is also the defining characteristic of the Commissioners' Plan for Manhattan: On a Cartesian plain, each point is uniquely defined as the intersection of two numbered coordinates at right angles to each other. Exactly the same is true of the commissioners' New York: "5th and 42nd" and "8th and 125th" are the coordinates of two specific locations in Manhattan, and they provide all one needs to know to find them. In fact, any given point in the New York grid, just as in a Cartesian coordinate system, can be defined as the intersection of two numbered rectilinear lines. The three distinguished commissioners, aided by Randel's conscientious survey, effectively transformed the island of Manhattan into a Cartesian mathematical plain.

The Battle for Manhattan

Why did the commissioners choose to make New York into an abstract mathematical landscape? Certainly not because they saw no alternatives. Morris had lived for years overseas and had traveled to all the great European capitals and many other cities. He was intimately familiar with Paris, where royal, revolutionary, and later imperial administrations were working to carve out great boulevards from the tangle of medieval streets, a project culminating with the work of Baron Haussmann half a century later.[44] All of the commissioners, furthermore, had been to Washington, where L'Enfant's dream of a republican capital, with its broad avenues converging on star-shaped plazas, was slowly taking shape. And in the commissioners' final report, written by Morris, he acknowledges that they considered whether "they should adopt some of those supposed improvements by circles, ovals, and stars, which certainly embellish a plan."[45]

The question becomes even more acute if we consider that the Commissioners' Plan, from its very inception, had few friends. The landowning families that had dominated New York since the days of New Amsterdam were dead set against it—hardly surprising as the plan ran roughshod over their farmland and elegant estates, the foundation of their power and prestige. In 1818, even as Randel was still traversing the island, producing his ninety-two detailed maps, they found their voice in a six-page pamphlet entitled *A Plain Statement Addressed to the Proprietors of Real Estate in the City and County of New York*. The author, who identified himself solely as "A Landholder," freely admitted that he had a personal interest in arresting the spread of the grid but pleaded a higher motivation: As a result of the near-dictatorial powers given to the commissioners, he warned, "we live under a tyranny, with respects for the rights of property," such as "no monarch in Europe would dare to exercise."[46] "Every man who has an interest in the prosperity and happiness of the community to which he belongs; who is susceptible of indignation at the thought of private property being invaded by public authority, without necessity and without compensation," he proclaimed, should stand with him.[47]

It did not take long for the identity of the anonymous "Landholder" to be publicly revealed: he was Clement Clarke Moore, a professor of classical literature and divinity at the Episcopalian General Theological Seminary (which he had helped found) and the reputed author of the all-time Christmas favorite, "'Twas the Night before Christmas."[48] More to the point, Moore was what he claimed, a New York landholder, and his Chelsea home was threatened with annihilation by the advancing grid. If nothing was done, 8th and 9th Avenues would soon slice through the bucolic estate, and 19th through 23rd streets would complete its metamorphosis into an urban checkerboard of uniform rectangular blocks. As anyone who has visited the Chelsea neighborhood in New York can attest, this indeed is precisely what happened. But in 1818, Moore still believed that the fate of Chelsea was not sealed and that a concerted protest by men such as himself might yet stem the gridded tide.

Whether Moore was truly the one who has enchanted generations of children with his tale of Santa Claus's midnight visit has never been fully established. But *A Plain Statement*, whose authorship is not in question, shows that without a doubt, Moore knew how to put words to powerful use. Moving on from his complaints of tyranny and the waste of public funds, he reserved his sharpest barbs for the doleful aesthetics of the new New York. The commissioners, he charges, had an aversion to any bump in the ground and "seem resolved to spare nothing that bears the semblance of a rising ground." These were men, he continues, "who would have cut down the seven hills of Rome, on which are erected her triumphant monuments of beauty and magnificence, and have thrown them into the Tyber or the Pomptine marshes."[49] The commissioners' propensities for "levelling and filling," according to Moore, "are lamented by persons of taste, as destructive to the greatest beauties of which our city is susceptible."[50] As for those who "still think the beauty of the city improved by its being reduced to a uniform flat surface," he adds acidly, "they already have ample room to gratify their love for plane surfaces." Hudson's Square was already level, and the area from "Spring-street to Greenwich-lane" would soon be reduced to a "dead flat."[51]

Moore, in the end, gave up the fight and became a wealthy man by selling off his estate, block by rectangular block.[52] Nevertheless, his prediction that "persons of taste" would object to the "uniform flat surface" of the commissioners' Manhattan proved prophetic. Over the coming decades, a steady barrage of insults would be hurled at the unpopular grid by the cultural trendsetters of the age. We have already seen what the fashionable *Athenaeum* of London had to say about the "frightful" vision of the grid.[53] Others were even less complimentary. Edgar Allan Poe spent the spring of 1844 in New York and bemoaned the fate of its beautiful old mansions, soon to fall victim to the advancing grid. "In fact," he wrote, "these magnificent places are doomed. . . . In some thirty years . . . the whole island will be desecrated by buildings of brick, with portentous facades of brownstone."[54]

Five years later, Walt Whitman (1819–92), still a young man but already well-known as an essayist, editor, and poet on the rise, denounced the erasure of New York's natural landscape, lamenting that "greater favor is not given to the natural hills and slopes of the ground." Instead we have "our perpetual dead flat, and streets cutting each other at right angles," which "are certainly the last things in the world consistent with beauty of situation."[55] Writing in the same year, as revolutions raged across Europe, William A. Duer, the longtime president of Columbia College, saw a danger greater than depressing ugliness in the uniform grid. "The levelling system," he warned, "reduced the superficial aspect of the city to an equality corresponding with the political condition of its inhabitants." That he regraded such egalitarianism as the opposite of a virtue was made clear by what followed: "In this process, not only that variety and undulation of surface, which contributed both to its health and beauty were destroyed, but the scythe of equality moved over the island."[56] In Duer's imagination, the equality of blocks might soon lead to the equality of the guillotine.

The twentieth century brought no respite from the steady stream of denunciations of the grid by taste arbitrators of their times. In 1906, Henry James called it New York's "primal topographic curse," an "inconceivably bourgeois scheme" that was the result of "the uncorrected labor of minds with no imagination."[57] Lewis Mumford, the twentieth century's foremost American critic of urban architecture, denounced the "long monotonous streets that terminated nowhere," and an aging Edith Wharton recalled the "cramped horizontal gridiron" New York of her youth, "hide-bound in its deadly uniformity of mean ugliness."[58]

From its very beginning, enmity to the grid was widespread, and by midcentury it encompassed everyone who should have mattered, including New York's powerful old families and its cultural elite. Why, then, did the commissioners choose such a radical plan for the city, one that would effectively erase all that existed on Manhattan? And why did they replace it with a rigid, uniform, and boundless mathematical grid, covering the entire island? Over the years, the question has exercised both historians and urban critics, but in the absence of records from the commissioners' deliberations, no clear answer has emerged. One popular explanation is that the plan was designed to assist in land speculation, to the immense benefit of New York's wealthiest citizens. Typical of this line of thought is John W. Reps, author of the most comprehensive history of urban America, who ends his account of New York with a devastating critique of Morris and his colleagues. "One cannot avoid the conclusion," Reps writes, "that the Commissioners, in fixing upon their plan, were motivated mainly by narrow consideration of economic gain." For him there can be only one justification for this rigid and seemingly unimaginative plan: "As an aid to speculation, the Commissioners' plan is perhaps unequaled."[59]

There is, to be sure, an inherent plausibility to Reps's position. For there is no denying that the progress of the grid over Manhattan was accompanied by a frenzy of land speculation and that great fortunes were made in this way, most famously by the former fur trader John Jacob Astor.[60] And yet as an explanation for the commissioners' choice it falls markedly short. For the fact is that the people who were most likely to benefit from the sale of Manhattan lands were dead set against the plan. These were the landholders of Manhattan, men like Clement Clarke Moore who were also the city's social and economic elite and would be expected to have the commissioners' ear. Yet as Moore's pamphlet shows, and as the century-long stalemate in city planning that preceded the commissioners' appointment clearly indicates, these men's main concern was for preserving their existing lands and the autonomy of their private family estates. Nothing could be more antithetical to their interests as they saw them than a plan to carve up the island, their estates included, into uniform blocks. As for men like Astor, who did indeed grow fabulously rich by speculating on Manhattan's rectangular plots, they were newcomers who had no standing when Morris and his fellow commissioners made their decisions. Their wealth, and their consequent influence, may have followed the grid, but it did not initiate it.

Others, meanwhile, have argued that the grid, which transformed an island dominated by the large estates of the few into one in which "all blocks are created equal," is the embodiment of republican egalitarian ideals.[61] This view too seems plausible: the contrast between the uniform squares of the great republican metropolis and great converging boulevards of the royal capitals of Europe is too stark to be considered a mere coincidence. And yet the commissioners who designed the grid were landowners of old family and very far from being egalitarian reformers. Their leader, Morris, in particular, was famous for his contempt for the masses, complaining at one point of "the insolent familiarity of the vulgar" and predicting that democracy is but one agonized step on the way to monarchy.[62] Such men were unlikely to be standard-bearers for the equality of man.

More recently, New York historian Gerard Koeppel has argued that the great grid we know today is, in fact, an accident. By late 1810, Koeppel argues, the commissioners had hardly begun the work of designing the street plan of Manhattan. With time running out on their mandate, they latched on to the existing grid of the Common Lands, surveyed by Casimir Goerck in the 1780s and 1790s. Koeppel even proposes a date when this momentous decision took place: November 29, 1810, just four months before the commissioners' appointment was due to expire. The grid, Koeppel suggests, far from being the product of an enlightened vision, "is an excuse for a plan, arrived at with little thinking and with time running out, based conveniently on an existing plan already on the ground, the Goerck Common Lands grid of 1796."[63]

Yet this view too has its drawbacks. For one thing, from 1808 to 1810 Randel spent much of his time measuring crosslines from river to river, some of them extensions of Goerck's cross streets and others parallel to them.[64] Koeppel accounts for these measurements as an effort to locate the Common Lands on the island, but it is hard to see why that would be a priority unless those lands were to play a central role in the commissioners' vision for the island. And with Randel repeatedly laying lines precisely where the commissioners would ultimately lay down streets, it seems beyond unlikely that this is entirely a coincidence. Then there is Randel's map of the Albany Post Road, which shows the Manhattan grid in great detail in the summer of 1810.[65] This effectively disproves Koeppel's timeline, which has the commissioners settling on the plan only in November of that year. Finally, it stretches credulity that Morris and his colleagues, appointed to map out a street plan for New York, simply ignored their charge until three months before their term was due and then slapped something together on short notice, like a student late on an assignment. The grid plan, as the commissioners surely knew, was not an easy solution but a wholesale assault on both the natural topography and the social fabric of Manhattan, and they would hardly have foisted it, with little thought, on generations of New Yorkers. It is far more likely that the grid plan was a serious possibility for several years before the commissioners submitted their report and that it guided Randel as he conducted his survey.

Why Manhattan Was Gridded

But if the grid was not a real estate speculation scheme, not a monument for democratic ideals, and not a slapdash last-minute plan, why did the commissioners choose it? The obvious place to start resolving this question is to examine what the commissioners themselves had to say on the matter in the only place they directly addressed it: in the *Remarks of the Commissioners for Laying Out Streets and Roads in the City of New York, under the Act of April 3, 1807,* which was submitted as the commissioners' final report, alongside the map prepared by Randel. The *Remarks,* as they came to be known, have most often served as "Exhibit A" for the prosecution, by those who claimed that the Commissioners' Plan "fitted nothing but a quick parceling of the land, a quick conversion of farmstead into real estate, and a quick sale."[66] Indeed, there is no denying that the text has a strong utilitarian bent and says nothing about higher ideals that may have been involved in the planning process. This is less than surprising considering that Morris, who almost certainly wrote the report, was not given to soaring flights of rhetoric. Thomas Jefferson liked to present even his pragmatic policies in the guise of eternal and universal ideals; Gouverneur Morris, in contrast, presented even the most idealistic triumphs, including the US Constitution, as down-to-earth solutions to practical problems. And so it was with the commissioners' *Remarks.*

A closer reading of the text, however, reveals hints of something else: if not grand Jeffersonian ideals, then at the very least some solid republican principles. A case in point is the passage, early on in the *Remarks,* that is most often quoted to illustrate the crass materialism of Morris and his colleagues. Before setting on their task, the report relates, the commissioners considered "whether they should adopt some of those supposed improvements by circles, ovals, and stars, which certainly embellish a plan, whatever may be their effect as to convenience and utility." After pondering the matter, however, they decided against such elaborate designs. "They could not but bear in mind that a city is composed principally of the habitations of men, and that straight-sided and right-angled houses are the most cheap to build and the most convenient to live in. The effect of these plain and simple reflections," they conclude, "was decisive."[67]

The commissioners' emphasis on cheapness and convenience certainly justifies the view that they aimed at a utilitarian solution to the problem of New York streets. But this only brings up the question of why they would choose utility over other design considerations. All three commissioners, after all, were familiar with L'Enfant's plan for the national capital, which set in stone the republic's constitutional power structure. And Morris, at least, was intimately familiar with the great European cities where grand boulevards, open plazas, and monumental buildings all reflected the glory of the monarch. Indeed, when he referred to "circles, ovals, and stars," it seems well-nigh certain that he had these royal capitals in mind.

To understand why Morris dismissed the European template of cities that are both bustling and elegant in favor of a "cheap and convenient" grid, it is worth examining another of his works, in which he contrasted European and American ways of life. In 1806, just one year before his appointment as New York street commissioner, Morris published an essay entitled *Notes on the United States of America*, purported to be an account of the young nation addressed to a European friend.[68] Morris, as he makes clear, is no enemy of European monarchy and considers it much preferable to democratic rule by "mob," "scum," and "designing scoundrels."[69] He even considers it likely that the United States will ultimately end up a monarchy. Nevertheless, he insists, there are important difference between Europeans and Americans: In Europe, according to Morris, a man is judged by "his stars, his ribbands, his military commission or noble descent." America, in contrast, is a meritocracy: "He who behaves himself well will be well received," regardless of fancy uniforms or ancient family. "He will be estimated at what he is worth, and if he has merit," and all for a simple reason: in America, "those who wander from the path of industry will soon be entangled in want." Europeans, says Morris, have the luxury of judging people by external trappings; Americans, in contrast, must judge people by their true worth if they are not to perish.[70]

The parallels between Morris's *Notes on the United States* and the commissioners' *Remarks* he authored five years later are too striking to be coincidental. Here the stars, ribbands, and noble descent of the *Notes* have become "circles, ovals, and stars"; both are emblems of European fluff, and both are simply unsuitable to the conditions and national character of Americans. The cities of America have no use for pretty but empty symbols of status and refined taste. They are to be as unassuming and practical as the people who inhabit them. The American people are unpretentious and industrious, necessarily so if they are to avoid want; American cities must therefore be equally unpretentious and facilitate American industriousness.

All of which goes a long way in explaining the commissioners' choice of a uniform grid as their plan for Manhattan. The city, as Morris and his colleagues perceived it, was not meant to make a grand ideological statement, like L'Enfant's Washington; nor was it meant as a brilliant display of urban elegance, as Paris would soon become thanks to the renovation work of Baron Haussmann. Whether it would be beautiful or ugly, imposing or simply boring, mattered little. Its purpose was different: to serve as the grounds on which industrious and enterprising Americans would pursue their ambitions and create their own fortunes, to their own benefit and ultimately to that of the young nation. In itself, the street plan of the city was nothing, just a blank slate on which Americans would build their own destiny. As much as possible, the island of Manhattan, with its hills, farms, and villages, was to become a featureless, empty, space, in which nothing preexisted and anything was possible.

Morris's and his colleagues' views on the principles that should govern American cities also come through in a different section of the *Remarks*, where the commissioners explain why they abandoned their earlier intention to accommodate existing street patterns where they existed. Initially, they explain, they sought to "amalgamate" their plan with streets already in place, so as "not to make any important changes in their disposition." This, they continue, was desirable because it would make their plan both more acceptable to the public and cheaper. In the end, however, they had to abandon the effort. Accommodating all existing street patterns was simply impossible, while accommodating some and not others would be blatantly unfair. For those who object to the commissioners' high-handed disregard of existing streets and roads, Morris has an answer: "It will perhaps be more satisfactory to each person who may feel aggrieved to ask himself whether his sensations would not have been even more unpleasant had his favorite plans been sacrificed to preserve those of a more fortunate neighbor."[71]

All of which is to say that in imposing their own plan and overruling existing ones the commissioners acted out of principle: no individuals should be preferred to others, and it is better, therefore, to disregard the interests of all people than to prefer the interests of some over others. There may be echoes here of Jefferson's grand assertion that "all men are created equal," but they are distant and partial echoes, to be sure. Morris and his colleagues had no intention of challenging the existing social order in New York or the dominance of property holders within it. But they do show great concern for the principle of fair play and a desire to make sure that all New Yorkers have the opportunity to work and succeed. The attitude is very much in line with Morris's assertion that in America, unlike in Europe, people are judged by their work and industry, not by "ribbands and stars." In gridded New York, equal, unremarkable, and practical blocks will give preference to no man but will ensure that all men have a chance to prove their worth.

It was not only Morris, among the commissioners, who expressed such views. His colleague Simeon De Witt wrote much the same two years after completing New York's street plan, stating that Americans are a uniquely industrious "inventive people."[72] America, De Witt continued, provides uniquely favorable conditions for exercising this inventiveness, for it "leaves leisure to the mind to wander through the mysterious, unfathomable repositories of *possible things*; to the boundless field of improvement before us."[73] America, in other words, is a clear field of possibility for its enterprising citizens, in which they will invest their energy and creativity.

Morris would no doubt have agreed. Even Jefferson, who sought to turn the western United States into a blank slate in which settlers could build their own lives and the new nation, could not have said it better. And De Witt leaves no doubt as to the key tool that would make America into the wondrous land of opportunity it was destined to be: "Perspective... sprung from the Science of Mathematics" would make it so. Where most men see nothing but chaos, the man trained in mathematical perspective "sees the distinct myriad of parts, wonderfully formed and put together by infinite wisdom to constitute a whole, perfect in all varieties of proportion, shape, color, and purpose." Perspective, in other words, transforms the chaotic space of untrained perception into an orderly mathematical space in which men can take action. It becomes, in De Witt's words, "the illimitable field which a beneficent Providence has opened up for the activity of man."[74]

All of which brings us back to the commissioners' decision to adopt a uniform grid pattern for the streets of the future New York. Both Morris and De Witt (we have no record of Rutherfurd's views) believed that Americans were uniquely industrious and inventive, more so than any people in Europe. They believed that unlike Europeans, who were defined by external trappings of ancestry and status, it was Americans' destiny to make use of their talents to pursue their own success and prosperity. America itself, it followed, must be the land in which industriousness and inventiveness, these notably American traits, would be offered free rein, for the benefit of its citizens and of the nation as a whole. As much as possible, the land must be free of natural obstacles and free of human historical limitations, an empty virgin land in which enterprising men pursue their goals unhindered. It should be blank and empty so as not to constrain human action, but it must also, as De Witt insists, be organized and coherent, or else all work will be for naught. Most of all, it must bespeak of infinite possibility.

A blank and empty space that is also coherent and organized and that allows for any and all action can only mean one thing: an abstract, uniform, and featureless mathematical space, where nothing preexists and everything is possible. This, the commissioners decided, would be the space of the greatest American city, the kind of space that would permit it to grow and thrive in a uniquely American way, so different from its European counterparts. To designate it as such, the commissioners chose the one pattern that defined mathematical space—its blankness, emptiness, and uniformity, as well as its unbending order: the universal Cartesian grid.

In settling on a grid pattern for Manhattan, the commissioners were not acting out of narrow self-interest. All three, after all, were landholders—Morris and Rutherford among the greatest—and by choosing the unrelenting grid, they were acting against their own interests and those of their class. They were not representing the interests of real estate speculators either: when they received their commission, the local politics of New York were dominated by landowners such as themselves, not by speculators, and they had no reason to prefer these particular interests to any others. Unsurprisingly, there is no record that they did. Finally, they were not acting to make New York into a monument for egalitarian democracy by designing a city composed of a multitude of identical blocks. The contempt in which their leader, Morris, held the common multitude makes it clear that nothing was further from their mind.[75]

What drove the commissioners was neither narrow commercial interest nor high-minded idealism. It was, rather, their deep-set beliefs about what America is and consequently what its greatest metropolis should be. America is the land where free men can exercise their talents and pursue their interests, unencumbered by tradition, rank, or external constraints. New York was the place to make it all possible. A flattened island of numbered streets and avenues, all straight and all intersecting at right angles, is a blank slate. It is an island with no past, no history, no hierarchy, and no landscape, an island in which every point is like any other, defined simply by two numbered coordinates. Most importantly, it is an island in which every point can be made into anything, grand or petty, by the industriousness and imagination of its owner or tenant. This is what the commissioners sought, and this was indeed what they created. Many over the past two centuries, have denounced the grid as oppressive, ugly, unimaginative, and all in all a pathetic excuse for urban planning. The commissioners would not have cared in the least: for them it was the perfect plan for the city in which American enterprise, energy, and innovation would be given free rein. And no one from their day to ours, not even the grid's fiercest critics, can claim they did not succeed.

The Founding Brothers

In December 1809, even as John Randel was crisscrossing Manhattan while getting sued by its most respectable citizens, his boss, Gouverneur Morris, got married. The news was a stunning blow to relatives of the fifty-seven-year-old confirmed bachelor, who had been looking forward confidently to the day they would inherit their portion of his considerable fortune. But now that Morris had a wife, and a few years later a son as well, his inheritance seemed secure, and his disappointed relatives could only watch despairingly as their chance of striking it rich petered away. Just as surprising as the marriage itself, however, was the identity of the bride: for the man who throughout his life had been intimately involved with aristocratic ladies and society women had married his housekeeper.

Yet "Nancy," as the new Mrs. Morris was known, was no ordinary housekeeper. Born Ann Carry Randolph, she was the scion of one of the grandest and wealthiest families in all of Virginia. In 1792, at the age of eighteen, she was living on a plantation called Bizarre with her sister Judith and Judith's husband, Richard Randolph, who was also a cousin of the two. When slaves on the plantation discovered the dead body of a white baby, Nancy and her brother-in-law were accused of adultery, and Richard was criminally charged with the baby's murder. The evidence seemed damning, but Richard was nevertheless acquitted. This was in part thanks to the able defense of his lawyers, Founding Father Patrick Henry and future chief justice John Marshall, but even more thanks to the fact that slaves were forbidden to testify against whites. As a result, Richard went back to the plantation, where he continued to live with his wife and his sister-in-law, who was, or at least likely had been—his mistress and partner in crime.

Life in Bizarre could not have been happy, and things did not improve when Richard died suddenly a few years later. Nancy, whose name was attached to adultery and possibly murder, was considered unmarriageable and consequently lived as a family dependent. In 1805, she was finally asked to leave, and, lacking a reliable income, she drifted northward in much-reduced circumstances. When Morris, who had known her as a young girl, heard of her situation, he offered her a position as his housekeeper, which she readily accepted. In April 1809, she began working at his estate in Morrisiana. By the end of the year, they were married.[76]

The marriage of Gouverneur and Nancy Morris lasted only until his death seven years later, but it was by all accounts a happy one. It also had an unexpected side effect, for it made Morris a relative of one of his greatest political adversaries, a man whose vision for the Union was a polar opposite of his own and with whom he had been trading barbs for a full three decades: Thomas Jefferson. It could hardly have been otherwise, for the great Virginia families were closely related and often intermarried. Nancy, as it happens, was the third cousin, as well as close friend, of Jefferson's daughter Martha. Even more significant was the fact that her brother Thomas Mann Randolph was Martha's husband. Consequently, when Gouverneur married Nancy, the old rivals, Morris and Jefferson, became in-laws.

This, in some ways, is pure happenstance, the result of the unusual personal circumstances of the woman who connected them. But it also reaffirms that, despite their rivalry and profound differences, Morris and Jefferson were never personal enemies and that, in fact, as founders of a new nation, they shared more than meets the eye. The contrast between the two men's visions for America are, to be sure, inescapable: Jefferson was a convinced democrat and founder of the Democratic Republicans, who believed in (though did not always practice) the equality of man, despised hierarchies, hated cities, suspected commerce and "speculators," and insisted that free yeomen famers in the West will form the backbone of the republic. As minister to France, he consorted with radicals and reformers, and when revolution came, he wholeheartedly and enthusiastically supported it. Morris, in contrast, was a committed Federalist, believed social hierarchies to be both necessary and desirable, was deeply suspicious of the masses, and considered democracy a form of mob rule and a stepping stone on the way to autocracy. He believed that cities and their commerce were the engines of the nation's economy, as well as centers of culture and refinement, and that western settlers were ignorant and uncouth. As Jefferson's successor in Paris, he consorted with royalists and aristocrats, did his best to save the lives of the king and queen, and considered the notion of a French republic a joke.[77]

And yet, if one looks beneath the surface, it suddenly seems that what Morris and Jefferson held in common far outweighed what set them apart. Most fundamentally, both believed that the United States was a new kind of nation, radically different from the great European monarchies. Frenchmen, Englishmen, Spaniards, and Prussians are born into a society in which their lives and expectations are circumscribed by class, by law, and by traditions going back hundreds if not thousands of years. Americans alone are unencumbered by ancient tradition and face an open world full of opportunity and promise. For both Morris and Jefferson, America is an open field of activity, a blank slate on which its citizens will pursue their own goals unhindered, a blank slate on which its free citizens will inscribe their own stories and thereby build a nation.

Jefferson and Morris held very different views on how and where the new Americans should expend their boundless energies. Jefferson believed that the future of the republic lay in the independent yeomen farmers who would settle the vast western expanses; Morris believed that the future of the young nation lay in the strength of its commerce and finances, concentrated in the Union's great cities. Yet remarkably, in seeking to make the United States into the country they knew it must become, they came up with precisely the same solution: transform the rich natural landscape of America into a featureless abstract space. Mountains, valleys, forests, and streams would be transformed into a featureless, homogenous, and empty mathematical space, marked by a boundless, regular, and uniform Cartesian grid.

Jefferson applied it to the open landscape of the West; Morris applied it to the streets of New York and by extension to all the American cities that followed and replicated the pattern. Jefferson, more than anyone, is responsible for the relentless grid that marks rural America; Morris, more than anyone, is responsible for the ubiquitous grid that marks its cities. Together, they created the unique pattern, so familiar as to be almost invisible, that distinguishes America from any country on Earth: a vast, homogenous, and monotonous Cartesian grid that encompasses the continent, from its vast open spaces to its densest urban hubs.

The Great American Grid, wrote Camillo Sitte, the late nineteenth-century Austrian urban critic, reflected the fact that "America lacked a past, had no history, and did not yet signify anything else in the civilization of mankind but so many square miles of land."[78] Sitte did not mean this as a compliment. Such packing of people "like herrings in a barrel," he goes on to explain, might be appropriate for uncultured outposts like America, but certainly not for civilized Europe. And yet Jefferson and Morris, both staunch American patriots, would for the most part have agreed with Sitte. Like him, they saw America as a land without history, which signifies nothing in itself but a vast open and empty space. And like him, they believed that a limitless and open Cartesian grid encompassing the land signified precisely that. But this is where their similarities end: Sitte saw the grid and lamented the absence of an American past; Morris and Jefferson saw the same and hailed the promise of an American future.

6

···········

Anti-geometry

The Juggernaut

In 1811, when the commissioners submitted their plan, the southern limits of the projected grid were still far north of the outer limits of the settled city. But within a few short years, the city not only caught up with the grid but began expanding rapidly northward. By 1824, 5th Avenue, already the trendsetting artery of the city, was opened up to 13th Street; four years later, it went up to 24th Street; and by 1837, it extended all the way to 42nd Street.[1] Advancing at a similar pace farther to the west, 9th Avenue had reached 30th street by 1830, and 3rd Avenue was not far behind.[2] Nothing, it seemed, could stand in the way of the forward-marching grid, not even the few breaks in the pattern put in place by the commissioners themselves. In the most drastic departure from the original plan, the Parade Grounds, the largest open space in the projected city, stretching from 3rd to 7th Avenue and 23rd to 34th Street, was unceremoniously discarded in 1829 and disappeared beneath the regular pattern of rectilinear streets.

The grid took over not only country estates, like Clement Clarke Moore's Chelsea, but soon entire villages as well. First to go was Greenwich, in 1825, and later on the villages of Bloomingdale, Manhattanville, and Harlem.[3] On the east side, the Post Road, one of the island's main north–south arteries, soon disappeared from the map, though its counterpart on the west side fared better. Here the rocky terrain forced a slower pace of development, and the Bloomingdale Road, granted repeated reprieves over the years, was ultimately incorporated into the city plan. Known today as Broadway, it snakes through rectilinear Manhattan, intersecting the avenues and streets at odd angles and providing the most notable relief from rectilinear uniformity.

The pattern for the grid's northward march was established early. The first to arrive when a street or an avenue was to be opened were the surveyors, measuring and marking its future path and recording the contours of the terrain. Next came city officials, expropriating the land destined to become a street, assessing the impact of the development on property values, and levying a tax on the residents to pay for the "improvement." If a house happened to stand in the way of a future street (and according to a recent estimate, 40 percent of New York houses did), then it was either demolished or—as was the case with many of the finer homes—moved out of the way. Then came teams of workmen with shovels, pickaxes, and (later) explosives, blasting hills and filling in dales to make a level ground fit for a straight city street.[4]

Once the street was laid out, the work was still far from done, as it could take years before it became connected to the advancing city. In the meantime, it could be, as one diarist described Madison Avenue in 1867, "a rough and ragged track . . . and hardly a thoroughfare, rich in mudholes, goats, pigs, geese, and stramonium" and the occasional "Irish shanty."[5] Yet over time, year after year, as the grid marched northward, the city followed in its wake, turning the muddy paths into paved streets and avenues, straight, broad, and lined with houses. An 1850 map shows the city densely built from its southern tip to about 23rd Street, with sporadic development extending all the way to 50th Street. Twenty-nine years later, the city had advanced to 59th Street, and development on the east side, though not continuous, reached as far as 135th Street. By the 1890s, the grid was filled out to 155th Street, where the commissioners had set its northern boundary, and new plans were being laid out for the northern tip of the island.[6]

Decade by decade, the gridding of New York City proceeded. Generation after generation of officials, recorders, surveyors, and laborers did their work and passed from the scene, while the project itself continued on its course, sweeping all before it. "The time will come," wrote one perceptive observer in 1858, "when New York will be built-up, when all the grading and filling will be done, when the picturesquely varied rock formation of the island will have been converted into the foundations for rows of monotonous straight streets, and piles of erect angular buildings."[7]

So indeed it appeared in the mid-nineteenth century, when the ultimate victory of the grid over all obstacles, natural and human, seemed only a matter of time. And yet it was the author of these words who did more, perhaps, than anyone to limit the grid's extent and impact. His name was Frederick Law Olmsted, who in later years would grow famous as the founder of the American school of landscape architecture, and whose naturalistic style became a hallmark of American design. In the 1850s, however, he was just a junior member of a group of well-connected New Yorkers determined to hold the grid at bay and counteract what they viewed as its nefarious effects.

The men who stood up to oppose the forward march of rectilinear New York in the 1840s and 1850s were a different breed from those who had protested the commissioners' rectilinear plan in previous years. Back when Randel was crisscrossing the island with his surveyors, he was pelted with vegetables, chased off properties, and sued repeatedly by landowners intent on protecting their property rights and fending off the streets and avenues they knew would follow in his wake. Some years later, gentlemen like Clement Clarke Moore tried to preserve their estates by rousing popular opinion against the advancing grid. In contrast, the men who took over the opposition to the grid in the middle decades of the century had no property that might be affected and no financial stake in the design of urban Manhattan. They were, rather, New York's and America's arbiters of refined taste, its cultural and intellectual trendsetters, who spoke with the authority of a confident and unchallenged elite. When they took a stand against the advancing grid and pronounced on its evils, they could not be accused of pursuing their narrow self-interest at the expense of the general public. Rather, they spoke as the guardians of the public good, whose duty it was to warn their fellow citizens of the dangers inherent in the ongoing transformation of their city. As a result, their opposition to the grid proved far more powerful than the rearguard actions of their predecessors and far more effective than any lawsuit. Even more impressively, the remedy they offered carried the day just when it seemed that all of Manhattan would soon become a gray checkerboard of uniform rectangular blocks.

It was, without a doubt, a star-studded group, whose names read like a who's who of not only New York's cultural elite but nineteenth-century America's literary luminaries. One prominent member was Washington Irving (1783–1859), the United States' first literary sensation. Back in 1807, as a young and irreverent essayist, Irving and his fellow "Salmagundi" celebrated the chaotic New York that they knew by contrasting it with its rival, the rectilinear metropolis of Philadelphia. The Philadelphians, in the view of Irving and his fellows, lived in a city "as fair, and square, and regular, and right-angled, as any mechanical genius could possibly have made it." This, they opined, had made Philadelphians an "honest, worthy, square, good-looking, well-meaning, regular, uniform, straight-forward, clock-work, clear-headed, one-like-another" kind of people who "walk mathematically, never turn but at right angles, think syllogistically, and pun theoretically." Meanwhile, "the people of New York—God help them—tossed about over hills and dales, through lanes and alley, crooked streets—continually mounting and descending, turning and twisting—whisking off at tangents, and left angled triangles, just like their own queer, odd, topsyturvy rantipole city, are the most irregular, crazy-headed, quicksilver, eccentric, whim-whamsical set of mortals that ever were jumbled together . . . and are the very antipodeans to the Philadelphians."[8] Decades later, when New York was fast becoming the very definition of a rectilinear metropolis, Irving was eager to salvage what was left of his beloved eccentric city.

Another member of the anti-grid coalition was James Fenimore Cooper (1789–1851), Irving's only rival for American literary acclaim, who railed against those who "scourge the very 'arth with their axes" and complained that "lots" had become the sole topic of conversation in New York society.[9] Yet another, as we have seen, was the young Walt Whitman, bemoaning the ugliness of New York's "perpetual dead flat" and rectilinear street plan.[10] But the undisputed leaders in the campaign to limit the New York grid were two men who are no longer household names, though they were famous enough in their day. One was William Cullen Bryant (1794–1878), a poet of national acclaim, as well as the longtime editor of the *New York Evening Post*. "Commerce is devouring inch by inch the coast of the island," he warned on the pages of his journal in 1844, and if any part of it was to be saved, "it must be done now."[11] The other was Andrew Jackson Downing (1815–52), a landscape theorist, designer, and popular writer who was also the editor of the *Horticulturalist* magazine. Rectilinear New York, he wrote in an 1852 editorial, was nothing but an "arid desert of business and dissipation."[12] Appalled by the seemingly unstoppable grid that was devouring Manhattan block after block, Bryant and Downing joined forces in a campaign to save what was possible from being swallowed up by that relentless juggernaut.[13]

City without a Soul

Why did the shapers of opinion in New York and elsewhere turn so decisively against the grid in the middle decades of the nineteenth century? The plan, after all, had by this time been in place for nearly four decades, and the disgruntled property owners who had tried to derail it in its early years had been silenced or, more often, bought off. The required work of surveying, leveling, and opening streets was passed on from one city administration to the next in seamless succession, as the grid took over acre after acre of Manhattan. Whether one liked it or not, the imposition of the rectilinear grid on the landscape of Manhattan had become routine, and the future layout of the city seemed preordained. Why, then, the sudden surge of principled opposition from the most distinguished members of the country's cultural elite?

Part of the reason, no doubt, was the impact of the Romantic movement that was sweeping both Europe and the United States. Romanticism idealized simplicity and a connection to nature, prized rural over urban landscapes and untrammeled nature over both. Indeed, there could hardly be anything less romantic than the uniform city grid that was systematically overtaking Manhattan. Yet there is more to the tide of opposition to the New York grid than simply the ebb and flow of cultural trends, especially since, by this time, European Romanticism was well past its prime. What was it about Romantic sensibilities, one may wonder, that midcentury American intellectuals found so alluring?

A possible answer may lie in the fact that the middle years of the nineteenth century were a challenging time for the young republic, especially so for its old elites. Wave upon wave of European immigrants were crashing on American shores, bringing with them different faiths, customs, and languages, threatening to overwhelm the country's established population and subvert its cultural norms. Nowhere was this threat felt more keenly than in New York, where most of these new immigrants landed and where many set up their own ethnic enclaves. Meanwhile, as the guardians of taste and morals watched in horror, self-made men such as real estate speculator John Jacob Astor, with little culture or education, amassed vast fortunes that threatened to make a mockery of the power and influence of New York's old families. As a German immigrant himself and the very embodiment of runaway capitalism, Astor was everything they most feared and detested.

No one expressed the distress of New York's old elite better than Downing. "The city," he complained, "is always full. Its steady population of 500,000 souls, is always there; always on the increase. Every ship brings a live cargo from over-peopled Europe, to fill up its crowded lodging houses; every steamer brings hundreds of strangers to fill its thronged thoroughfares. Crowded hotels, crowded streets, hot summers, business pursued till it becomes a game of excitement, pleasure followed till its votaries are exhausted."[14] With boatloads of immigrants regularly arriving in the city and flooding its streets and unbridled capitalism running unchecked in the commercial districts, men such as Downing and Bryant saw a city that had lost its soul. The greatest metropolis in America, they feared, was quickly becoming a cesspool of violence, greed, and corruption.

Meanwhile, troubling news was trickling in from across the ocean. The popular revolutions of 1848–49 were welcomed with enthusiasm by Americans, who thrilled at the sight of common people overthrowing tyrannical monarchs and establishing representative governments. It seemed to Americans that the people of Europe, at long last, were following in the footsteps of their own revolution decades before. And so when the early triumphs of what became known as the "Spring of the Peoples" were checked and the revolutions summarily defeated by the forces of autocracy and reaction, the news was greeted with profound shock. The ideals America stood for, far from marching to victory, faced defeat on every side. Add to that the specter of sectional strife, which was tearing the nation apart over the issue of slavery, and the overall prospects for America seemed doubtful indeed.[15]

In the mid-nineteenth century, the United States was facing nothing less than a crisis of confidence. Back in 1776, the republic was founded on the promise that it would provide justice, equality, and peace and serve as a beacon for the world. Yet three-quarters of a century later, despite impressive gains in territory, wealth, and population, these promises remained unfulfilled, and their attainment seemed a distant and receding dream. The nation's growing cities were filling with what seemed to genteel Americans like the dregs of the earth, who could never be expected to understand, much less participate in, democratic political life. The promise of equality was mocked by the reality of slavery, making all Americans vulnerable to the charge of blatant hypocrisy. The conviction that free people seeking their own betterment would live in harmony did nothing to protect the republic from the looming clouds of civil war. And with the defeat of the European revolutions, the United States remained what it had been since its founding—lonely and isolated, the sole republic among the world's major nations. Whether a nation, to use Lincoln's words, "conceived in liberty" and a government "of the people, by the people, for the people" would ultimately survive or perish seemed very much an open question.[16]

It was in these troubling times when the no-longer-young United States faced adverse winds and an uncertain future that the opposition to the Manhattan grid coalesced. And little wonder, for the grid was a child of the confident early years of the republic. Along with the Jeffersonian grid that was carving the West into precisely measured squares, the Manhattan grid was turning the American landscape into a blank slate, an empty space of infinite possibility. In gridded America, free enterprising men, liberated from the constraints of both human tradition and natural landscape, would pursue their interests and thereby create a free, prosperous, and peaceful nation. America so conceived was a manmade land and would thrive because of it. The grid was the expression of this boundless optimism, standing for the unshakable belief that free men let loose in an empty mathematical space will build the perfect society and ideal nation.

Yet in the gloomy decades that led up to the Civil War, such optimism seemed not only misguided but dangerous. The idea that people let loose in a world of their own creation would pursue the path of virtue and thrive seemed refuted at every turn. The expansion westward, which Jefferson believed would produce a new breed of freedom-loving and self-reliant people who would form the backbone of the republic, now seemed to portend the spread of the scourge of slavery and oppression. In the cities, meanwhile, the dense urban neighborhoods of Manhattan seemed to beget the opposite of republican virtue. Whether it was the unchecked greed of speculators or violence and moral decrepitude among the poor and newly arrived, the manufactured world of a city such as New York was the breeding ground for the worst that humanity could offer. The great gridded landscape of the United States, both rural and urban, was designed to give free rein to men's best impulses and thereby create a utopian society. Instead it became a trap, an artificial world of human corruption from which there was no escape. If the republic and its people were to be saved, the grid's genteel enemies concluded, the grid must somehow be contained.

But how does one contain this juggernaut that had been systematically and irresistibly sweeping through Manhattan, year after year and decade after decade, like a divine scourge? Arresting it in its track seemed all but impossible, as the momentum of the multigenerational project would easily overwhelm any obstacle thrown in its path. So Bryant, Downing, and their fellows offered an alternative. The grid, they conceded, would ultimately expand and cover the entire island, but at its heart they would plant an antidote: a vast anti-grid at the very center of the city in the form of an expansive, green, and naturalistic public park. Such a park, Downing explained, would elevate the people of the city out of their sad condition and raise them to a higher level of civilization. New York's "half a million of people," he declared, "have a *right* to ask for the 'greatest happiness' of parks and pleasure grounds."[17]

Bryant, Downing, and their fellow anti-grid activists were confident that a public park, or at least the right kind of public park, would go a long way toward mitigating the deleterious effects of the grid. The problem with the grid, as they saw it, was that it created a hermetically sealed artificial world, fully insulated from nature. The grid, after all, was a perfect blank slate readied for human activity, which resolutely ignored or suppressed all external constraints. Morris and his fellow commissioners may have believed that this empty Newtonian landscape would bring out the best in Americans. Uniform and unresisting, it would provide an ideal space for economic enterprise and the creation of a dynamic and virtuous nation. But forty years later, Bryant and Downing knew better. City dwellers left to their own devices in an artificial landscape were not the heroic founders of a new nation but lost souls, victimized by their own fears, unrestrained passions, and unchecked greed.

The only hope for the inhabitants of the artificial urban world, the only escape from the trap of the city grid, was to get back in touch with natural open space. Out in the countryside, it is the regular rhythms of nature that reign supreme, not the caprices of men. There men and women are reminded that they are part of a larger whole and that nature is greater and more powerful than anything humans can devise. When viewed against the grandeur of the natural world, human ambitions seem small, even petty. Nature is, in effect, the perfect antidote to the evils of the gridded city: the grid breeds pride and the unchecked pursuit of ambition and gratification; nature, in contrast, teaches humility, restraint, and refinement, putting ambition and greed in their proper place. To save the city from itself, nature must be brought into its very heart.[18]

The Green Hills of England

Bringing nature into the bustling gridded streets of Manhattan seemed like an impossible proposition, but the anti-grid campaigners believed they had a model for just such an endeavor: the public parks of England. For whereas French gardens were famous for their geometrical formalism, English gardens were renowned for creating an idealized natural landscape, which was precisely what Downing and Bryant were looking for. In 1850, Downing traveled to England and spent much of his time touring gentlemanly estates, private gardens, and, most significantly, public parks. These last were relative newcomers, as great gardens had until recently been cultivated exclusively on the private estates of the aristocracy, safe from the prying eyes and trammeling feet of lowly commoners. But beginning in the early decades of the nineteenth century, several royal gardens in London had been opened to the public and become St. James's Park, Kensington Gardens, and Hyde Park, among others. Soon new parks were being created in English cities, purposefully designed for the general public. It was one of these—Regent's Park in London—that most impressed Downing.[19]

While the American urban reformers hoped to use the English gardens as a counterweight to the grid, the English style itself was originally a reaction to a very different kind of geometrical landscaping: the brilliant and imposing formalism of Versailles. Back in the seventeenth and early eighteenth centuries, the French style, as perfected by Le Nôtre, set the standard for gardening across Europe, and England was no exception. Just like his royal cousins on the Continent, the newly restored Charles II sought to create his own royal garden in Greenwich Park, and in 1662, he even hired the French master to design it. In the event, Charles II was a poor relative to Louis XIV, and Greenwich Park a poor imitation of Versailles. The garden did more to highlight the yawning gap in wealth and power between the two monarchs than it did to enhance the English king's prestige.[20] Even so, Charles's admiration for the formal French style was undiminished. Not long afterward, he added Hampton Court and St. James's Palace to his complement of French-style gardens, thereby setting an example for all the great nobles of England. Over the following half century, every grand estate in the land, it seemed, was replanted in the French style, complete with star-shaped plazas, broad arrow-straight allées, and grand manors at the focal point where all lines meet.[21]

Not everyone, however, was pleased with this French invasion of England, even if it took the pleasant form of tree-lined avenues and brilliantly colored parterres. And little wonder, for the French formal garden was the very embodiment of the ideals of absolute monarchy, ideals that patriotic Englishmen viewed with a mix of apprehension and revulsion. Versailles, the purest and most glamorous incarnation of the French gardening style, was a geometrical world in which every pool, stone, tree, and blade of grass had its precisely assigned place, presided over by the royal palace. Anyone walking the paths of such a garden was instructed in the hierarchies of the world, as immutable and universal as geometry itself, and the unchallengeable power of the monarch who ruled them. It was a message that was enormously appealing to the great royal houses of Europe, and monarchs from Madrid to Vienna to St. Petersburg rushed to build their own French gardens to rival Versailles. But it was also a message that was bound to raise fierce opposition in London and throughout England.[22]

The reason is that in the late 1600s England was just emerging from a bitter civil war brought about in large part by the pretensions of its monarchs to absolute rule. Earlier in the century, James I and his son Charles I had wished to follow the example of their counterparts across the Channel and reign without parliamentary consent or constraint. What followed was twenty years of bloody conflict in which parliamentarians fought royalists, different factions fought one another, and the king, captured by his enemies, was publicly beheaded. Ultimately, in 1660, fearing social upheaval, Parliament felt compelled to recall the son of the unfortunate Charles I and restore the monarchy. But this did not mean the parliamentarians were willing to cede to Charles II the powers his forebears had aspired to. Quite the opposite. They were determined that the monarchy would remain strictly limited and that true power would reside in the hands of Parliament. When James II, who succeeded his brother on the throne, ignored these demands, converted to Catholicism, and allied himself with Louis XIV, he was overthrown and replaced by the more amenable William and Mary. And when the Stuart line finally expired in 1714, it was succeeded by the Hanoverians, who never challenged parliamentary supremacy and reigned contentedly as limited constitutional monarchs.

All of which is to say that in the decades around 1700, nothing was sure to inflame English opinion so much as royal pretensions to absolute rule. And as nothing presented those claims so clearly and powerfully as French formal gardens, it seemed all but predestined that they would come under attack. Early hints of this opposition could be found as early as the 1650s and 1660s, when John Milton, poet and dedicated anti-monarchist, was composing *Paradise Lost*. Tellingly, in the age of Le Nôtre, when the French formal style set the standard for garden design across Europe, Milton's ultimate garden—the Garden of Eden—was nothing of the sort. Paradise, for Milton, far from being the regimented geometrical space of Versailles, was "a sylvan scene," where "cedar and pine, and fir, and branching palm" ascended "shade above shade, a woody theatre [o]f stateliest view." Similarly, the flowers of Paradise did not reside in artificial "beds and curious knots" but "nature boon, poured forth profuse on hill and dale and plain."[23]

By the early decades of the eighteenth century, Milton was joined by a steady stream of criticism and mockery of the formal style of gardening in the English press. In China, reported the essayist and politician Joseph Addison (1672–1719) in 1712, "the inhabitants . . . laugh at the plantations of our Europeans, which are laid out by the rule and line." As for himself, he continued, he was very much in agreement with the Chinese perspective: "I would rather look upon a tree in all its luxuriancy of boughs and branches, than when it is thus cut and trimmed into a mathematical figure," he wrote. Furthermore, he "cannot but fancy that an orchard in flower looks infinitely more delightful than all the little labyrinths of the most finished parterre."[24] His friend, the poet Alexander Pope (1688–1744), fully agreed: "We seem to make it our study to recede from nature," he complained the following year, turning "the various tonsure of greens into the most regular and formal shapes."[25]

Critiquing the formal French style was all well and good, but what would replace it? Milton described the garden of Eden as "natural," and Addison contrasted nature favorably with the artificial garden, saying that "there is generally in nature something more grand and august than what we meet in the curiosities of art."[26] Pope did them one better by creating at his estate in Twickenham one of the earliest naturalistic gardens seen in England. Even so, well into the 1740s, the French formal style, inspired by Le Nôtre, remained the norm among English lords seeking to beautify their estates. It was around that time that a young landscaper named Lancelot Brown began to make the rounds of English country houses, offering to transform their surrounding parks into something very different. At every stop, he would survey the terrain with its proprietor and assure him the estate had great "capability." Over the next few decades, "Capability" Brown, as he came to be known, would establish the naturalistic English gardening style and transform the landscape of England.

Born around 1715, Brown was the son of a yeoman farmer from Northumberland and was apprenticed to a gardener in his youth. Moving south in his early twenties, Brown found a position as an assistant to William Kent, one of the leading garden designers of the time, at Stowe, Lord Cobham's lavish estate near the town of Buckingham. When Kent died in 1741, Brown succeeded him, and over the next decade, he worked to make the park at Stowe into the magnificent

The Serpentine at Stowe, Buckinghamshire, England. From iStock.com/John Entwistle.

natural-style garden it remains today. Gone were the broad arrow-straight avenues leading up to the great house; gone were the star-shaped plaza where they intersect, the elaborate and colorful parterres, the elliptical and octagonal ponds, and the precisely trimmed trees places at regular intervals. Gone, in other words, was everything that made Stowe Park a French geometrical garden. In their place, Brown introduced a large serpentine lake at the bottom of a "Grecian Valley," with bubbling brooks snaking their way toward it through the brush. There were no broad avenues in Brown's designs, no straight lines, and no right angles, but only paths that wound their way up hills and down valleys, along streams and through shady woods and thickets, emerging occasionally into the open where broad vistas beckoned. It was, and remains, a mild and beautiful landscape that invites visitors to immerse themselves in the illusion of a gentle and soothing nature.[27]

For illusion it undoubtedly was. Brown's gardens, in contrast to Le Nôtre's, were designed to look and feel artless, as if created by nature herself without the intervention of a human hand. In fact, not only were they carefully and precisely designed and engineered but creating and maintaining them required a workforce that would have been worthy of the master of Versailles. To excavate the Grecian Valley at Stowe, Brown had his workmen remove 23,500 cubic yards of soil and rock, using nothing but shovels and wheelbarrows. At Blenheim Palace, he employed a small army of workers to dig up two massive lakes and then dammed the river Glynne to fill them up. At Heveningham in Suffolk, he raised a lawn by twelve feet. On occasion he was obliged to remove entire villages from an estate, evicting or relocating their inhabitants much as Le Nôtre had done at Versailles. And to preserve his vast parklands and prevent them from reverting into a giant bog, he designed elaborate drainage systems, which lay hidden beneath the surface of lawns and woods. Brown's gardens were designed to look natural, but the key term here is *designed*." If "natural" means letting nature do its work, then there was nothing natural about them.[28]

Yet Brown and his designs were all the rage among the English aristocracy. Visitors from far and wide came to admire the sylvan beauty of Stowe and invited its designer to transform their own estates. Even while still employed by Lord Cobham, Brown began advising his neighbors, who were eager to create their own magical gardens, and when his patron died in 1749, he set up shop as an independent landscape designer. Commissions began pouring in immediately, as everybody who was anybody in England needed a Brown-designed naturalistic garden to verify their membership in the elite. To appreciate Brown's touch today, one need only pay a visit to what is likely his largest and most spectacular project—the transformation of the grounds of Blenheim Palace, residence of the Dukes of Marlborough. Originally designed by John Vanbrugh in the French style, the park was transformed by Brown into an entrancing naturalistic landscape. At Blenheim, where the palace can be viewed through the thicket of a shady wood, floating above a peaceful lagoon, one can still catch a glimpse of Brown's original vision. Two and a half centuries later, his magic is still there.

Blenheim Palace, Oxfordshire, England.

In three decades of working independently, Brown designed and transformed an astonishing 170 parks and gardens, the vast majority belonging to the landed aristocracy. Once, when offered a vast sum to landscape the estate of an Irish lord, he declined, on the grounds that "he had not yet finished England." Brown can be forgiven the boast, for with so many estates and gardens shaped by his hand and his vision, he had indeed transformed the English countryside. The popular image of England as a land of rolling hills, gentle valleys, and green meadows separated by streams and woods is in no small part the creation of Capability Brown.

The spread and popularity of Brown's designs did not stop at the borders of England. Across the Channel, Voltaire sang their praises, and Jean-Jacques Rousseau, the prophet of natural living, warned that if things continue as they are, "people will want nothing in their gardens that is found in the countryside."[29] Even Marie Antoinette, Louis XVI's discontented queen, insisted on having her own natural garden at Versailles. To the horror of royal traditionalists, a portion of the most iconic formal gardens in the world was carved out and landscaped as a simple meadow and pastoral village, where the queen and her ladies-in-waiting would play the part of milkmaids and shepherdesses. Americans were impressed as well. Thomas Jefferson spent much of the spring of 1786 touring English estates in the company of John Adams and his wife, Abigail. The Francophile Jefferson was no friend of England, but he could not restrain his enthusiasm for what he had seen. "The gardening of that country," he wrote in a letter to his Virginian friend John Page, "is the article in which it surpasses all the earth."[30]

What all these admirers of English gardening had in common—with the partial exception of the queen of France—was a deep suspicion of the royal absolutism embodied in Versailles. Voltaire criticized its excesses, and Rousseau derided its sterile formalism and superficiality. Jefferson, of course, was a republican who despised crowned heads, and in particular those who aspired to absolute rule. Even Marie Antoinette, though a queen herself, was driven to distraction by the formalities of life at court and longed for relief from the rigid hierarchies that defined French absolutism. And this indeed was what the English garden provided: a counterweight to the absolutist claims of the great continental geometrical gardens, and most particularly Versailles.[31]

Much like Versailles, an iconic English garden was a microcosm of the world and everything in it. Hills and valleys, meadows and forests, grasses, flowers, and trees, as well as rivers and lakes—as miniature oceans—all had their place on the grounds of Stowe or Blenheim. Yet when it came to the ordering of this world, the gardens of Le Nôtre, on the one hand, and Capability Brown, on the other, could not have been more different. At Versailles and elsewhere, Le Nôtre created an orderly world where everything had its eternally fixed place and all could be monitored from the palace. Order was manifest, hierarchy incontestable, and there was no escaping the all-encompassing royal gaze. Brown's designs too were far from chaotic—the peaceful lagoons surrounded by meadows and woods spoke of a deep universal order put in place by a benevolent God. Yet in contrast to Le Nôtre's gardens, that order is hidden and cannot be preknown or prejudged by any man. The only way to discover the true order of creation, Brown's gardens imply, is to wander its paths, go down the occasional blind alley, lose one's way, and then find it again. Bit by bit, over time, the outlines of an overall plan might emerge, providing a glimpse of God's design.

From this, the political implications followed. A visitor wandering in geometrical Versailles is instructed in a world in which order is eternally fixed and hierarchy is rigid and unyielding. The absolute rule of the monarch in his palace is as incontestable as Euclidean geometry itself, an integral part of an unchallengeable universal order. But if the same visitor then crossed the Channel and walked the trails of Stowe or Blenheim they would enter an altogether different world—hierarchical, to be sure, as the presence of the great house inevitably reminds any visitor, but ordered in ways that are mysterious and largely unknown. To uncover this world, and one's place within it, one must proceed not like an all-knowing geometer but like an experimental philosopher—slowly, carefully, through trial and error. In such a world, the authority of the master of the palace is far from a given and far from absolute, and its true extent is always in question. And unlike at Versailles, where all is exposed to the all-seeing royal gaze, there is much in an English garden that remains hidden from the great house, sheltered from the power of its master.

Le Nôtre's geometrical gardens are the embodiment of absolute monarchy and its inescapable royal gaze. Brown's anti-geometrical landscapes speak of limited authority, a hidden and never fully known order, and individuals whose place in the universal scheme is fluid and subject to perpetual negotiation. Versailles speaks the language of Louis XIV's royal absolutism; Blenheim speaks of limited monarchy and individual autonomy. The rejection of geometrical gardens in England, and the adoption of a naturalistic nongeometric style, was not only an aesthetic choice but a political stance. The rigid and all-encompassing hierarchies of Versailles, it stated clearly, will never be allowed to take root in England.

Capability Brown in his day had worked exclusively in the service of the great landholders of England, thereby recasting the countryside in the image of his naturalistic dreamlands. Working in gray dusty cities such as London or Liverpool would probably have seemed to him absurd, the very antithesis of the world he was working to create. It was not until the 1830s, more than four decades after his death in 1789, that things began to change. First in London, then in other cities, parks were opened for use not by a select few aristocrats but by the general public, including working men and women. Unlike Blenheim and Stowe, these parks were located in the midst of or adjacent to bustling urban spaces, providing a startling contrast to the nearby streets. In the 1830s, the first public parks opened in London, and by the 1840s, a municipality like Merseyside, across the river from Liverpool, could purchase land and develop a park for the express purpose of public recreation. And yet for all their novelty, when it came to design these new parks might as well have been located on the grounds of the rural estate of an ancient noble family. Public parks they may have been, but to their designers and users, they were, above all, English parks, which meant they must conform to Capability Brown's dreamy rural landscape. And so, from Regent's Park in London to Birkenhead Park in Merseyside, they were designed as large naturalistic gardens of green hills, wooded valleys, and bubbling streams.

A Refuge in the City

It is not hard to understand the attractiveness of English public parks to Bryant, Downing, and their fellow reformers. In style and design, developments like Regent's Park and Birkenhead were the direct successors of the naturalistic dreamscapes Brown had designed for England's great aristocratic houses. Their location, however, placed them at the heart of England's newly industrializing cities, a green counterpoint to the gray rows of factories belching black smoke into the skies. In both regards, of style and location, the American urban reformers believed that the English parks offered an answer to the ills afflicting Manhattan. To save the people of New York from the all-consuming grid that was devouring the city, they believed, nature must be brought into the heart of Manhattan, just as it had already been brought into London and Liverpool. A large and public English garden in the midst of the city would offer New Yorkers a refuge from the all-consuming grid and a chance to rediscover themselves amid the calming rhythms of the natural world.

Paradoxically, the use the Americans hoped to make of English gardens was the opposite of that for which they were originally designed. To Capability Brown and his contemporaries, the naturalistic style was meant to counter the rigid regimentation of Versailles. The formal geometrical gardens perfected by Le Nôtre represented the hierarchical and eternally unchanging world of French absolutism, which the English denounced as oppressive and tyrannical. The winding trails, irregularly shaped lakes, and charming hidden enclaves were meant to soften the sharp lines and rigid hierarchies of the French style, presenting a world in which order and hierarchy were more flexible and not as clearly delineated.

But in Manhattan, Bryant and Downing were confronting a landscape very different from the Euclidean symmetries of Versailles. To them, the problem with the uniform grid of Manhattan was not that it established an overly rigid hierarchical order; it was, rather, that it established no order at all. Like the great Jeffersonian grid of the western United States, the Manhattan grid erased the human and natural order that preceded it but offered nothing to replace it. Like the empty space of analytic geometry, it established a blank slate of infinite possibility to the people inhabiting it. The problem for the urban critics of the mid-nineteenth century was that this freedom from constraint did not bring about the peace, prosperity, and harmonious living that earlier generations of American had hoped for. To the contrary, in their opinion it created a human trap, where every sort of vice flourished. Enclosed in a completely artificial world and with no escape possible, the people of New York fell victim to the worst of human passions.

Yet remarkably, the solution the American reformers offered to the moral crisis of the masses was identical to the one English constitutionalists had used to counter the claims of royal absolutism: a renewed emphasis on the rhythms and order of nature, as embodied in large naturalistic gardens. To the Englishmen, the mysteries and irregularities of the natural world were a powerful antidote to royal claims of absolute knowledge and absolute power, expressed through a rigid geometrical order. To Americans, the same mysteries were a source of reassurance that a grand universal order, greater and more powerful than anything man could design, pervades everything in the natural and human world. Nature, as recreated in English-style gardens, was more flexible and irregular than the formal Euclidean gardens of Versailles, but it was also more regular and orderly than the chaotic free-for-all taking place on the streets of New York. Capability Brown and his colleagues relied on the irregularity of a world in which no lines were ever straight and no intersections were at right angles. Bryant, Downing, and their fellows emphasized the rhythms of nature, the deep regularities of the days and the seasons, the animals and the plants, which pervaded the world beyond the reach of human contrivance. Janus-like, the English garden embodied both freeform flexibility and deep unchanging order. The English and the Americans, confronted with contrasting challenges, each latched on to a different face.[32]

Armed with their newfound faith in the English gardening style, Bryant and Downing continued their fierce campaign for a public park in New York. By 1850, they seemed to have made some headway, as both mayoral candidates that year professed themselves true believers in the park idea and promised to make it a reality. When Ambrose C. Kingsland emerged victorious, he referred the matter to the notoriously ineffective Common Council to come up with a plan. Kingsland may have intended this as a way of avoiding the difficult and controversial decisions that a public park was bound to entail, all the while appearing to be true to his campaign promise. But the Common Council, under intense pressure from leading citizens, surprised all observers and came up with a practical solution: they recommended the purchase of 160 acres along the East River north of 66th street, an area known as Jones Wood. By the summer of 1851, the state legislature had approved the purchase, and New York seemed on its way to enjoying a sizable riverside park.

Except, however, that the park advocates would not have it. "One hundred and sixty acres of park for a city that will soon contain three-quarters of a million people?" asked Downing incredulously. "It is only a child's play-ground," he declared, noting that London had over 6,000 acres of "the grandest and most lovely park scenery" in the city or the surrounding suburbs. *"Five Hundred acres is the small-est area that should be reserved for the future wants of such a city"* as New York, he concluded.[33] With public opinion now firmly on their side, Downing and his fellow campaigners prevailed. By 1853, the Jones Wood plan was sidelined and ultimately dropped, and in its place the state legislature authorized the city to purchase the grounds from 59th to 106th Streets, between 5th and 8th Avenues, a narrow rectangle encompassing 777 acres.

Two years later, perhaps under pressure from real estate develop-ers and speculators, the Common Council had second thoughts and voted to chop off the southern end of the park, from 59th Street to 72nd. They may have believed that as the city moved closer to the park's boundaries they would be able to rid themselves of the rest of this troublesome obstacle in the path of the ever-expanding grid, but at any rate, they were disappointed. Mayor Fernando Wood, King-sland's successor, vetoed the Common Council's resolution, thereby preserving the park's boundaries. Then, in 1859, came the decision to expand the purchase area north to 110th Street, thereby establishing the park boundaries as we know them today. All in all, the area of "the central park," as it was then known, encompasses 843 acres, or just under 6 percent of the surface area of Manhattan.[34]

The park advocates had won a resounding and improbable vic-tory, but their leader, Downing, did not live to see it. On July 28, 1852, along with his wife and mother-in-law, he boarded the Hudson River steamer *Henry Clay* to travel from his hometown of Newburgh to Manhattan. Vying for the title of "fastest boat on the river," the *Henry Clay* was racing a rival when, just south of Yonkers, one of its boil-ers exploded and the ship caught fire. Eighty people died that day on the river, including a former mayor of New York and other members of the state's and city's elite. Andrew Jackson Downing, handsome, charming, successful, and only thirty-six years old, was among them. Had he lived, there is little doubt that Downing would have played a decisive role in the shaping of New York's Central Park. Already, up and down the Hudson Valley, he had designed and created dozens of private gardens in the same Romantic English style that he envisioned for Manhattan's urban park. And as the park's leading advocate, the job would have been his for the taking. As it happened, that position was left to a young acquaintance who admired Downing and shared his enthusiasm for all things natural and rural. He was, of course, Frederick Law Olmsted.

The Seeker

In 1850, at the very time that Downing was visiting London and ad-miring Regent's Park, Olmsted too was traveling the European coun-tryside both in England and on the Continent. At twenty-eight years old, Olmsted had impressed Downing in several articles he published in the *Horticulturalist*, and the two recognized each other as kindred spirits. When Olmsted returned from his travels the following year, Downing asked him to write a report for the *Horticulturalist* on his experiences in Germany. But Olmsted chose to write about something quite different: his visit to the new public park at Birkenhead, across the river from Liverpool. As far as Americans were concerned, the existence of the park was a well-kept secret. Downing himself visited Liverpool just a month after Olmsted had passed through, but despite his enthusiasm for English gardening, he made no mention of it and apparently had never heard of it. Olmsted himself heard about it by chance from a Merseyside baker, who begged him not to leave without seeing the new park. Olmsted did, and what he saw did much to shape his future career, not to mention the urban landscape of America.

"Five minutes of admiration, and a few more spent in studying the manner in which art had been employed to obtain from nature so much beauty, and I was ready to admit that in democratic America there was nothing to be thought as comparable with this People's Gar-den." So wrote Olmsted of his impression of Birkenhead in *Walks and Talks of an American Farmer in England*, an account of his travels pub-lished shortly after his return, which became his first literary success.

Walks and Talks is, for the most part, a sober and often critical description of rural England, but when it comes to Birkenhead, enthusiasm overflows: "We passed by winding paths," Olmsted continued, "over acres and acres, with a constantly varying surface" and "every variety of shrubs and flowers," all of it arranged "with more than natural grace." All in all, he concluded, "gardening had here reached a perfection that I had never before dreamed of."[35]

At Birkenhead, Olmsted became an instant convert to the concept of a naturalistic public park at the heart of the city. It was not only the aesthetics of the English garden that captivated him but the very idea of a genteel park in an urban setting, which struck him as profoundly democratic. "All this magnificent pleasure-ground," he exulted, "is entirely, unreservedly, and for ever the people's own. The poorest British peasant is as free to enjoy it in all its parts as the British queen."[36] This indeed was precisely what Bryant and Downing had been advocating for New York: a natural landscape accessible to all walks of life in the city, and in particular to the urban poor. It was only such a park, they believed, that could save the city's inhabitants from the moral and physical morass inflicted on them by the grid. At Birkenhead, Olmsted saw what such a park could be, and joined the campaign.

Frederick Law Olmsted has become so closely associated with the science and art of landscape design that it is easy to forget that he arrived at this vocation almost by accident. The son of a prosperous merchant from Hartford, Connecticut, Olmsted tried out numerous career paths in his youth: surveyor, mercantile clerk, and even common seaman on a ship in the China trade. He attended Yale but was forced to leave before completing his degree because of trouble with his eyes. When he finally settled on becoming a farmer, his father purchased a plot of land for him near Guilford, Connecticut, and when that failed to make a profit, he exchanged it for another farm near the south shore of Staten Island, in New York Harbor. Olmsted was an enthusiastic student of farming, both a reader and a contributor to journals such as Downing's. But he seemed far less enamored of the daily grind of farming, the unglamorous and often dull work that requires patience and perseverance but is essential for a farm's success. And so in 1850 he set out to Europe to study farming methods, but found himself drawn to public parks instead.[37]

Walks and Talks, published in 1852, proved more successful than expected, but that only complicated Olmsted's career prospects. The book sufficiently impressed the editor of the *New York Daily Times* that he offered to send him on a tour of the southern United States to report on conditions in the slave states. And so, instead of returning to his Staten Island farm, Olmsted embarked on a four-year journey through the American South that took him all the way from the Atlantic Seaboard to Texas. His dispatches, published regularly in the *Times*, were later published in several volumes and ultimately in an abridged collection entitled *Journeys and Explorations in the Cotton Kingdom*.[38] This made Olmsted a minor celebrity and, to northerners, something of an authority on conditions in the slave-holding states. But when his contract with the *Times* ended in 1857, Olmsted was thirty-five years old, still financially dependent on his aging father, and no closer to settling on a life path than he had been when he visited Birkenhead seven years before.

It was in that very year, however, that work on the creation of "the central park" in New York began in earnest. For several years already, the grounds of the future park had been carefully surveyed and mapped by Egbert Viele (1825–1902), who produced a drainage plan for the area and proposed a landscape design. But four years after the park boundaries had been approved and the ground purchased by the city, no ground had yet been broken and not a stone had been turned that would make Bryant and Downing's dream into reality. This, however, was about to change: In 1857, the state legislature appointed a board of commissioners of the park, which was aided by a star-studded advisory committee under the leadership of Washington Irving. The board then created the new position of park superintendent, an official responsible for clearing the grounds of existing structures and debris and preparing them for future landscaping.

Olmsted was once again at a career impasse, and so he decided to apply. "What else can I do for a living?" he asked his brother, by way of explanation.[39] He had no experience in public works management, but he was exceptionally well connected in New York, and that ultimately counted for more. With a letter of support signed by William Cullen Bryant, *New York Tribune* editor Horace Greeley, Harvard naturalist Asa Gray, and, most importantly, Washington Irving, he overcame the board's hesitancy and won the job. It turned out to be a felicitous choice. For Olmsted, quite apart from his enthusiasm for public parks, knowledge of farming, and qualifications as a journalist and writer, turned out to be an outstanding administrator.

This was fortunate because transforming the designated area into a park required nothing less. "A pestilential spot, where rank vegetation and miasmic odors taint every breath of air" is how Viele described it to the board of commissioners in an 1857 report.[40] He was, doubtless, exaggerating, but it was nevertheless true that the 777-acre (later 843-acre) tract was dotted with legal and illegal breweries, bone-boiling works, and hog farms, all draining into swamps and creeks that doubled as sewers. And while the landscape of the would-be park was forbidding enough, there were also the people who lived in it, who were equally the target of Viele's contempt. "Squatters, dwelling in rude huts . . . and living off the refuse of the city," he called them when reflecting on his work on the park years later. Even worse, he complained, "these people . . . were principally of foreign birth, with but very little knowledge of the English language, and with very little respect for the law."[41]

Viele's attitude was probably typical of the time. Far to the north of the dense city, the area of the future park seemed to old New Yorkers to be rough and untamed, well beyond the bounds of civilization. As for the tract's polyglot and multiracial inhabitants, they appeared to them a rough and unruly lot of suspiciously mixed descent. Black people, foreigners, and non-English speakers, they were exactly the kinds of people who sent chills up the spines of propertied New Yorkers, who were already anxious about the changing demographics of their city. Yet the reality was rather different. For the area that would become Central Park was far from the lawless no-man's-land that haunted the dreams of New York's respectable citizens.

According to modern estimates, about 1,600 people made their home within the boundaries of the future park, including many families. The vast majority of these were either recent Irish and German immigrants or members of Manhattan's growing African American community. And while they were, for the most part, poor compared to the propertied New Yorkers to the south, many were gainfully employed and far from destitute. They included day laborers and domestics but also skilled tradesmen such as tailors, carpenters, stonemasons, and even two innkeepers. Many paid rent to the absentee owners of the tract they lived on, while others built their homes on land whose ownership was unknown, technically justifying Viele's derisive description of them as "squatters."

The largest settlement in the future park was the African American hamlet known as Seneca Village, occupying the area between what are today 82nd and 86th Streets along Central Park West. Built on land purchased in the 1820s by its Black founders, it sported three churches and a population that remained remarkably stable over several generations. Many of its families owned the land on which they lived, and some had built sizable and even multistory homes. This made Seneca Village not only the most affluent community in the park-to-be but also the most fortunate. Because when the time arrived for the area to be cleared of its people, the roughly 20 percent who owned the land they lived on received compensation. The 80 percent who did not received nothing.[42]

The man responsible for cleansing the park of its people was, inevitably, Olmsted. Settling into his new position as park superintendent, he quickly recruited an army of a thousand men and sent them to clear the grounds of houses, works, shacks, and, most importantly, men, women, and children. When the park's people fought back, he organized his own police force called the "Park Keepers" and mercilessly evicted anyone who stood in his way. In Seneca Village, the three churches were demolished, the roomy houses razed, and the population dispersed. By the fall of 1857, the job was complete, the vacated land made ready for its future life as a city park. Olmsted was jubilant: "I have got the park into a capital discipline, a perfect system, working like a machine," he reported proudly to his father in early 1858.[43] But it is hard when considering the fate of the future park's residents not to recall the similar—if more brutal—fate of the native tribes who stood in the way of western expansion. The new America might be envisioned as an immaculate grid or as a natural garden, but either way, it had no room for people who did not fit its mold.

Meanwhile, as Olmsted was showing his skill in clearing and demolishing, members of the board of commissioners were having second thoughts about the existing plans for the park. Instead of following Viele's plan, which had been previously approved, in October 1857 they announced an open competition, inviting entrants to submit proposals for the park's design. Olmsted, who was fully engaged in his work as superintendent, gave no thought to entering the competition. But this changed when he was approached by an acquaintance, the English landscape designer Calvert Vaux (1824–95), who suggested that the two should collaborate and submit it as a joint proposal.

Vaux's invitation carried weight because he had been the business partner of the late Andrew Jackson Downing, the spiritual father of the New York Park. Back in 1850, when Downing was touring England, he met Vaux in London, was impressed by his skill and vision, and immediately offered him a job. That very year, Vaux picked up and moved to America, where he joined Downing in his design office in Newburgh, New York. Vaux continued running the office even after Downing's death, accepting and executing commissions throughout the Hudson Valley and beyond. In 1857, lured perhaps by the prospect of turning his late friend's dream of an urban park into reality, he moved to New York City and began campaigning with the commissioners to hold an open design competition for the park. The plea from the man who had been Downing's closest associate had its effect: the commissioners announced the competition, and Vaux, for his part, immediately approached Olmsted with an offer to join forces. He may have recognized early what the rest of the world would soon discover: that Olmsted was Downing's natural successor as the leading advocate and spokesman for naturalistic urban gardens.[44]

Faced with Vaux's offer, Olmsted voiced his reservations. Viele, whose design was being shunted aside, was the park's chief engineer and thereby Olmsted's immediate superior. If Olmsted was to enter the competition, it might be considered not only an unfriendly act toward Viele but rank insubordination. When asked, however, Viele indicated that he did not oppose Olmsted's plans. Perhaps he was being magnanimous, or perhaps he sensed that he was already on his way out and wanted to maintain cordial relations with a man who was clearly on the rise. Whatever the case may be, Olmsted hesitated no more and accepted Vaux's offer, and the two set about putting together a proposal that would best express their ideals and their vision for the park. It was a partnership that, over the coming decades, would reshape the American landscape no less than Capability Brown had transformed the English countryside.[45]

The Central Park

Things moved quickly after this. In the fall and winter of 1857–58, Olmsted and Vaux spent many hours walking together through the grounds of the future park. Olmsted, who, thanks to his work as superintendent, knew the terrain better than anyone, no doubt shared his knowledge, as well as his ideas, with his new partner. Vaux, for his part, likely shared his experience as a professional landscape designer in both England and America. Every evening, after completing a full day's work on the park grounds, Olmsted would head over to Vaux's home at 136 East 18th Street, and the two would work on their design. Visitors who came by were invited to assist in filling in the grass in the green meadows and lawns in their map. They named their plan Greensward, a name that evokes grassy rolling greens and is likely a tribute to Birkenhead Park, which Olmsted had years before described as a "green-sward."[46] On April 1, they submitted their design to the park commissioners; on April 28, Greensward was announced the winner from among more than thirty proposals. Olmsted was appointed architect in chief, a position that made him the park's "chief executive officer . . . by or through whom all work on the Park shall be executed."[47] He was, effectively, the park's dictator, Vaux was his

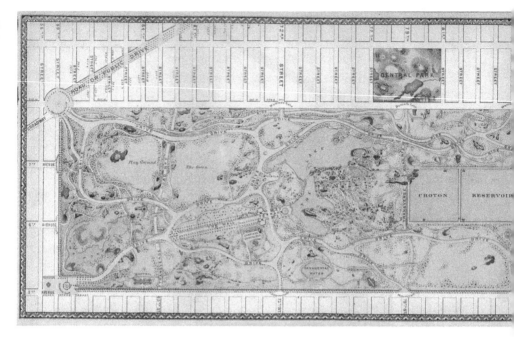

The Greensward plan.

deputy, and the two began immediately to make the vision of Greensward into reality.

What vision was it? At its core, Greensward aimed at countering the
rectilinear geometry of the city streets by bringing non-geometrical
nature into the heart of the city. That Olmsted was no friend of the
New York grid, or "the system of 1807," as he called it, should come
as no surprise. "There seems to be good authority," he relayed some
years later, "for the story that the system of 1807 was hit upon by
chance occurrence of a mason's sieve near a map of the ground to be
laid out." Morris and his colleagues, the story went, noticed the sieve,
placed it upon the map, and then asked, "'What do you want better
than that?' no one was able to answer."[48] The monotonous grid, in
other words, is a random creation, meaningless and pointless. The
fact that it was adopted at all can only be explained by the laziness of
the original commissioners, who lacked vision and imagination and
simply latched on to the simplest, most mechanical scheme available.

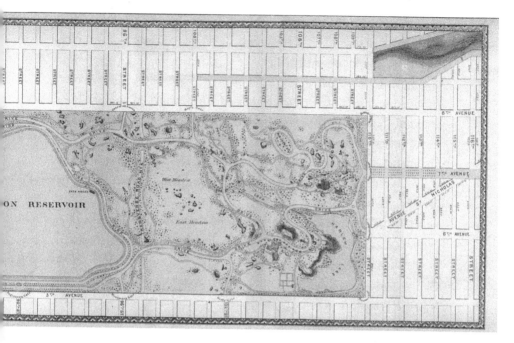

But what's done is done, Olmsted acknowledged, and all that remained was to limit the damage. This, he believed, meant surrounding the grid, both within and without, with lush naturalistic landscapes. Manhattan, a narrow island surrounded on all sides by growing cities (later boroughs), could not be encircled with a green band of parks and suburbs. It was not too late, however, to plant a park in an elongated rectangle that would soon become the heart of the bustling city. Even in its confined space, Olmsted and Vaux explained in their Greensward proposal, the park could still offer a decisive alternative, even a rebuke, to the grid. This is particularly true, they wrote, of the upper (i.e., northern) park, whose "horizon lines are bold and sweeping." This characteristic, they explained, "is the highest ideal that can be aimed at for a park under any circumstances," precisely because "it is in most decided contrast to the confined and formal lines of the city."[49] Thirteen years later, reporting to the Chicago South Park Commission, Olmsted insisted even more forcefully on the necessary contrast between park and grid: "The element of interest which undoubtedly should be placed first . . . in the park of any great city," he wrote, "is that of an antithesis to its bustling, paved, rectangular, walled-in streets." He also made clear the kind of park required to produce such a contrast: "a large meadowy ground, of an open, free, tranquil character." Sweeping horizons, greenery, rolling hills, changing landscapes, and curving paths winding their way through them are what a park had to be for Olmsted and Vaux. The park, in other words, would be everything that the gray uniformity of the geometrical city grid was not.

That such a park was essential to New York, as to any growing city, Olmsted had no doubt. Living in an unmitigated urban environment, he explained in an 1880 address to the American Social Science Association, breeds "vital exhaustion . . . nervous irritation, and constitutional depression," as well as tendencies that pass through "excessive materialism, to loss of faith, and lowness of spirit, by which life is made, to some, questionably worth the living." Only "the contemplation of beauty in natural scenery is practically of much value in alleviating these evils," he argued.[50] The purpose of an urban park, he wrote in an 1858 letter to the Board of Commissioners of the Central Park, is to bring these experiences into the city and make them available to those who would never otherwise know them. The park's "one great purpose," he wrote, is "to supply the hundreds of thousands of tired workers, who have no opportunity to spend their summers in the country, a specimen of God's handiwork that shall be to them, inexpensively, what a month or two in the White Mountains or the Adirondacks is, at great cost, to those in easier circumstances."[51] Such an escape from "the cramped, confined, and controlling circumstances of the streets of the town," he added some years later, provides its people with "a sense of enlarged freedom" that is essential for their "health, strength and morality."[52]

The grid, it seems, had come a long way since it was proposed by the New York Street Commissioners in the early years of the nineteenth century. To Gouverneur Morris, Simeon De Witt, and John Rutherfurd, the uniform grid signified freedom from the empty ostentation of European capitals and freedom from external constraint. A blank slate of uniform emptiness, it was the ideal landscape in which to unleash the energy and ingenuity of free men and women pursuing their interests and thereby building a nation. This landscape of possibility was the very essence of what the New World had to offer. But to Olmsted and Vaux, writing six decades later, this landscape of freedom had become a prison. Instead of encouraging optimism, energy, and enterprise, it bred disease and moral and physical decay. The gridded city, when conceived, had been the space where Americans would thrive; now, surviving its toxic environment required escape, regular doses of the open vistas of wild nature, and the green pastures of the countryside.

Thanks to Olmsted's position as park superintendent, he and his partner required no pause after winning the design competition and immediately began the work of creating a natural sanctuary at the heart of the gridded city. Transforming the rocky and swampy rectangle into the pastoral dreamlike landscape of Greensward was a public works project on a gigantic scale, but happily for Olmsted and his men, they did not need to rely exclusively on shovels and pickaxes to shape the terrain, as Capability Brown's men had a century before. Instead, they made ample use of explosives—a quarter million tons of gunpowder, in fact, by the time the park was opened to the public in 1873. During the same time, workers hauled 4,825,000 cubic yards of earth and stone (equivalent to 10,000,000 one-horse cartloads) in and out of the park and treated the barren original soil with over half a million cubic yards of fertilizer. They planted 17,300 trees in 1859 alone, and the number of plantings kept growing year by year, reaching 74,730 in 1862. To water this urban forest and the meadows surrounding it, the architect in chief and his men laid down twelve and a half miles of water pipe just in 1861. To drain it, they installed an additional 62 miles of earthenware pipes below the surface. Along the way, the Park Commission became one of the largest employers in New York City, and likely in the country. At the height of the work, in 1859, Olmsted had at his command more than 3,800 workers.[53]

Bit by bit, the park began to take shape. A large new reservoir was dug, irregularly shaped like a natural lake and familiar today to park visitors. Two older rectangular reservoirs remained in place for several decades until filled in and covered up by the Great Lawn in 1929. A Lake was created with shady coves hidden beneath green canopies and outstretched branches. The rough and varied terrain of the Ramble, seemingly wild but in fact meticulously designed and executed by Olmsted, was carved out between the lake's arms. Elegant dreamlike structures, as if of a bygone age, were erected one by one—the Bow Bridge, Bethesda Terrace, Belvedere Castle, and others. A grassy field was designed as a "parade ground" but soon came to be known as the Sheep Meadow after the herds of sheep (and, early on, deer) that grazed there. And the "promenade," a broad straight avenue leading north to Bethesda Terrace, represented the only formal element in the entire park. In one of Greensward's most creative strokes, Calvert and Vaux sank the four crosstown traffic arteries cutting through the park into channels eight feet deep, making them all but invisible from level ground. They then bridged them with overpasses, ensuring the smooth circulation of foot, horse, and carriage traffic through the park itself.

Creating Central Park was a massive undertaking of both manpower and money, involving blasting, leveling, digging, planting, and a considerable amount of new construction. Yet as Olmsted explained in an 1870 address to the American Social Science Association, it was well worth the effort and the cost. A city such as New York, he explained, is "noted for the frequency of certain crimes, the prevalence of certain diseases, and the shortness of life among its inhabitants." Even worse are the morals exhibited by "young men in knots ... rudely obstructing the sidewalks," harassing "men, women, and children whom they do not know," and entertaining themselves in "lighted basements" where they "see, hear, smell, drink, and eat all manner of vile things." The antidote for all that is a park, "a simple broad open space of clean greensward, with sufficient play of surface, and a sufficient number of trees about it to supply a variety of light and shade." There, Olmsted argues, the same young men, along with other city residents, "may stroll for an hour, seeing, hearing, and feeling noth-

The Ramble and the Lake in Central Park.
Photo courtesy of the author.

ing of the bustle and jar of the streets," and, "in effect, find the city put far away from them."[54]

There is no doubt, according to Olmsted, that Central Park, even in its unfinished form, was already having a laudable effect on the life of the city. "No one who has closely observed the conduct of the people who visit the Park," he declared, "can doubt that it exercises a distinctly harmonizing and refining influence upon the most un-fortunate and lawless classes of the city,—an influence favorable to courtesy, self-control, and temperance."[55] The cold, impersonal, rec-tilinear geometrical grid of New York was, to Olmsted, the root cause of the uniquely urban disease of moral degeneration. The smooth, soothing, natural, and nongeometrical contours of Central Park were the one and only cure.

More than a century and a half after Olmsted began his work on Central Park, we may not be as sanguine as he about its overall moral impact. Repeatedly over its history, the park has flourished then declined, and has endured changes and intrusions from the surrounding city. Some were relatively benign, in the form of car traffic, playing fields, statues, a skating rink. Others, such as the Hooverville encampment in the early years of the Depression, less so. At times it seemed that Olmsted's best hopes for the park, that it would offer a blissful Arcadia to escape the ills of the city, seemed fulfilled. At other times urban ills such as pollution, homelessness, and criminality found a comfortable home within the park's boundaries. Yet despite the changes wrought by a complex history, Central Park today is still easily recognizable as Olmsted and Vaux's vision, laid out in the Greensward plan. It is still now, as it was to them, a sliver of "God's handiwork" in the heart of the gridded city.

An American Microcosm

The layout of Manhattan as we know it today is a testament to deep tensions between competing visions of America and to a historical moment in which they came into conflict. One vision saw America as a land of limitless opportunity for free enterprising men, who would build their own lives—and thereby the nation. It was an ideal embraced and promoted by the United States' founders and early generations, who believed that unlike European nations, America can be molded into whatever form its inhabitants choose to make of it. The monotonous, uniform, all-pervasive grid, that graph-paper landscape made up of straight numbered lines that meet at right angles, is the embodiment of this vision. Whether on the great western planes or on the streets of New York, it erases both the natural and human features of the landscape, setting the stage for perfectly free and unconstrained human action.

The difficult decades of the mid-nineteenth century, however, brought a different vision to the fore. Alarmed at the sectional strife over slavery, the waves of foreign immigrants washing up on their shores, and the perceived threat to the social order, Americans grew weary of the grid and all it represented. A fully artificial landscape that gave free rein to human enterprise also let the worst of human vices thrive, unchecked by God, nature, or tradition. To counter the seemingly unstoppable march of the rectilinear world of the grid, they offered an alternative vision: rather than a land of infinite opportunity, America was for them a green Arcadia, a refuge from bustling civilization where men and women can reconnect with the simple rhythms of nature. Natural rather than manufactured and uncompromisingly nongeometrical where the grid was boldly geometrical, this America was meant to counter the worst effects of the rectilinear world of the grid and, if not erase it, at least limit its reach and impact. In time, this vision would shape everything from the emergence of national parks on the margins of the western grid to the romantic modern suburbs on the outskirts of gridded metropolises. But the earliest and most influential expression of this view was undoubtedly Central Park in New York.

With its geometrically gridded streets and its anti-geometrical park, Manhattan is a microcosm of the competing dreams that have shaped much of America. It is imprinted with the boundless optimism of the republic's early years, as well with the darker, more circumspect vision that prevailed in the run-up to the Civil War. Its topography was defined at the exact moment when the two visions clashed, thereby creating both the rectilinear city and its opposite—Central Park—a rebuke nestled in the grid's very heart.

To some extent, the Manhattan landscape and its contradictions are a living memorial to that particular historical moment when it was forged in the middle decades of the nineteenth century. But the story of the different visions of America and the tensions between them did not end on the gray streets of New York or the green canopies of Central Park: They are as real, vital, and active in today's United States as they were when Olmsted and Vaux entered their Greensward plan into competition. The manmade versus the natural, the grid versus the anti-grid, the geometrical versus the anti-geometrical; It is a conflict of two visions that have made America what it is. And nowhere is it more simply, clearly, and concisely presented than where it all began: on the streets of Manhattan Island.

A landscape in conflict: the city and the park.

Beyond Manhattan

New York, with its rectilinear pattern of numbered avenues and streets covering most of the island of Manhattan, is likely the most systematically gridded metropolis in the United States, if not the world. Yet among the cities of the Eastern Seaboard, New York's graph-paper pattern was the exception rather than the rule. Philadelphia, New York's rival in size and wealth in the early years of independence, was a famously gridded city, whose north–south streets (though not the east–west ones) were numbered. There were also several colonial towns, including New Haven, Connecticut, in the North, and Savannah, Georgia, in the South, whose rectilinear streets were named, not numbered. But unlike New York's open-ended grid, these towns (including Philadelphia) were designed with elegant square blocks, clear rectangular boundaries, and a well-defined center.[56] The federal capital of Washington was unique in its adoption of the grand European style, with its broad avenues converging on star-shaped plazas, while many other towns, most significantly Boston, grew up organically and possessed no regular pattern at all.

Yet as Americans began pouring over the Appalachian Mountains in the decades after independence and then moving methodically ever westward, the new cities they founded took on an increasingly distinctive look. In the West, the open-ended grid was king. First among those western cities was probably the town of Marietta, Ohio, founded in 1788 by the surveyor Rufus Putnam as both his home and base of operations. Initially built as an armed camp to resist Indian assaults, it was soon overlain with a regular rectilinear pattern. Other Ohio towns followed, including Cincinnati (1788) and the state capital of Columbus (1812), both laid out as open-ended grids, and the general pattern for western cities was set.

Even when town planners, inspired perhaps by the example of Washington, DC, tried to lay out cities in the grand European style, their efforts mostly came to naught. In 1807, Judge Augustus Brevoort Woodward proposed an elaborate honeycomb pattern for the streets of Detroit, with major boulevards converging on star-shaped plazas at the center of each hexagon. But even the initial enthusiasm for the design, and its official endorsement by the city council, could not hold the standard grid at bay for long. By 1817, Woodward's plan was abandoned, and the city was fast laying down streets along the familiar gridiron pattern. Only the single semicircle of Grand Circus Park in downtown Detroit remains today of Woodward's visionary design. Indianapolis, founded in 1821, was designed as a European-style capital, with avenues converging diagonally on a central square, home to the state government. But as the city grew and expanded, it quickly reverted to the standard rectilinear plan, leaving only a small portion of the old downtown to bear witness to the ambitions of the city's founders. In 1836, another midwestern capital, Madison, Wisconsin, was similarly designed, with avenues converging on a square state capitol, whose corners—not sides—were aligned with the points of the compass. But once the city expanded beyond the narrow neck separating Lakes Mendota and Monona, its streets adjusted their orientation, settling into the familiar east–west and north–south gridiron pattern. Whatever the designs of a city's founders, it seems, the pull of the grid proved irresistible.[57]

As the United States expanded and new cities were founded farther and farther to the west, the grid seemed to gain in confidence. If the fathers of Indianapolis and Madison still sought ways to avoid the gray uniformity of the grid and grant their cities a touch of European grandeur, the founders of cities farther to the west seemed to embrace the inevitable. Chicago was begun in 1830 as a small village on the shores of Lake Michigan. But as plans for a commercial canal moved forward, the village exploded within a few short years into a giant gridded metropolis, whose rectilinear streets soon covered an area far surpassing the size of Manhattan. Kansas City, Missouri, founded in 1853, settled into the standard gridiron pattern without a fuss, as did Denver, Colorado (1861), and Phoenix, Arizona (1867). San Francisco, California, was a small village in 1848, but the onset of the Gold Rush the following year made it the Gateway to the West for the hordes of prospectors pouring in. By 1850, the city had a population approaching thirty-five thousand, its steep hills covered by not one but two rectilinear grids, separated by the diagonal of Market Street. Further inland, the towns of Benicia (1847) and Vallejo (1850) were laid out more conventionally as single uniform grids. So was the future capital of Sacramento (1850), with the grounds of the state capitol formed by joining together several rectangular city blocks.

And so America was gridded bit by bit, ever more thoroughly and systematically as it expanded westward. The vast open prairies were divided into regular squares, marked by perfectly straight lines aligned with the points of the compass. The small towns and villages aligned themselves with their surroundings, creating their own rectilinear patterns. And the great cities too, despite early efforts at variation, ended up in a near-uniform checkerboard pattern. It is a unique pattern, a precise geometrical grid representing an abstract, empty, and uniform geometrical space. It is repeated on every scale and almost any terrain, in both urban and rural settings, and it is unparalleled anywhere in the world. It is a uniquely American way of perceiving, comprehending, and mastering space.

Enemy of the Grid

And yet the universal grid is not the only pattern that uniquely marks the American landscape and repeats itself time and time again. So is its antithesis, seemingly untouched "natural" spaces that are easily recognized by their irregular and curvilinear outlines, in which no two lines ever intersect at a right angle. If the Great Western Grid defines the land as the absolute space of analytic geometry, then the natural spaces are self-consciously and deliberately ungeometrical. This was the case in New York, where Central Park was designed as a slice of bucolic nature in the midst of the rectilinear city, an antidote to life in an artificial geometrical landscape. It was also, as it turned out, the beginning of a movement that would champion not only urban parks but also parklike residential suburbs and even wild national parks intended to preserve nature in all its pristine glory. Within a few short years, it would transform the American landscape.

The person who did more than anyone to spread the creed of the natural landscape was without a doubt Frederick Law Olmsted. In 1865, after years of bruising battles with the Central Park commissioners over the design and running of the park, he joined forces with Vaux once more for the design of a new park across East River in Brooklyn. The success of Prospect Park, as it is still known, soon led to further commissions across the country, from Buffalo, New York, to Knoxville, Tennessee, and from Bridgeport, Connecticut, to Milwaukee, Wisconsin. The specifics of the design varied from city to city: in Boston, for example, Olmsted designed the linear string of parks known as the Emerald Necklace; in Buffalo, he designed a series of parks throughout the city and connected them with broad green boulevards known as parkways. But in each and every case, the parks were landscaped much like Manhattan's Central Park, the grandfather of them all. They were large English-style naturalistic gardens, with varied terrain, plentiful greenery, water, and winding paths that stood in dramatic contrast to the bustling gridded streets all around.

The urban parks movement swept through the country in the later decades of the nineteenth and early twentieth centuries, and it was not long before every city from the Eastern Seaboard to San Francisco, California, was demanding a park of its own. Those who could not secure Olmsted's services hired others, but they too were required to design in the master's familiar style. Olmsted himself, meanwhile, was expanding his interests beyond parks to other types of urban designs. In 1866, during a sojourn in the West, he was hired to advise on the design of the College of California, better known today as the University of California, Berkeley. His proposal for an academic park, in which the buildings are nestled among greenery and winding paths, is the basis not only of the design of the famous campus but of much of the architecture and landscaping of American colleges to this day.

College campuses were a side interest for Olmsted, although twenty years after his proposal for Berkeley, he also assisted in designing its cross-bay rival, Stanford University.[58] The design of residential suburbs, however, was a more central focus and one of the key lines of business for the architectural firm he founded with Vaux. Olmsted was not the first American designer to create green parklike suburbs. In the early 1850s, the architect Alexander Davis, designed a leafy residential suburb in Llewellyn, and a few years later, David Hotchkiss was commissioned to design Lake Forest, north of Chicago. Like Olmsted, both architects were heavily influenced by the pastoral teachings of Andrew Jackson Downing, and their irregular curvilinear designs reflected that. But once again it was Olmsted's own design that became the standard by which all others were judged.

Riverside, Illinois, is today a Chicago suburb, but in 1868, when Olmsted was hired to design the new town, it was well outside the city's boundaries.[59] Olmsted's unique street plan is still clearly visible in the town, with long curving roads intersecting at odd angles. The housing tracts created in the spaces between the streets, far from being standard rectangular blocks, were curved and elongated ovals of various sizes and lengths, all of them nestled among wooded greenery. Viewing a similar Olmsted design some years later, an observer complained that the blocks are shaped "like melons, pears, and sweet potatoes" and one elongated block "like a banana." "It is a pretty fair park plan," the man commented with notable insight, "but condemns itself for a town."[60]

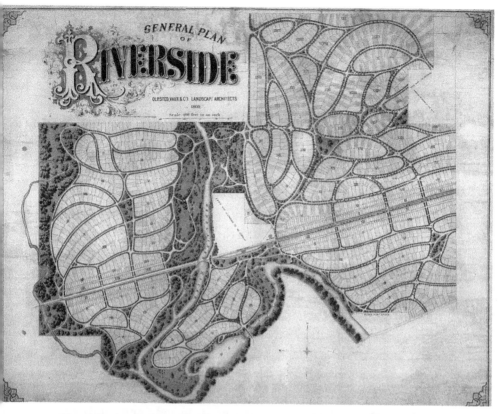

Olmsted's plan for Riverside, Illinois, 1869. From The NewYork Public Library,
https://www.nypl.org/research/research-catalog/bib/b20644637?originalUrl
=https%3A%2F%2Fcatalog.nypl.org%2Frecord%3Db20644637.

And yet a park design was precisely what Olmsted had in mind.
In fact, the purpose of his irregular design for the suburb was the
same as it had been for Central Park: to create the strongest possible
contrast to the gridded city. If a suburb is designed in this manner, he
explained in his report to the Riverside Improvement Company, "its
character will inevitably . . . be not only informal, but, in a moderate way, positively picturesque, and when contrasted with the constantly repeated right angles, straight lines, and flat surfaces which
characterize our large modern towns, thoroughly refreshing."[61] Just
as Central Park with its naturalistic landscape stood as a rebuke to the

surrounding New York grid, so would Riverside keep at bay the advancing rectilinear streets of Chicago. The one would subvert the grid from within; the other would strangle it from without. But whether a park or a suburb, the design and its purpose were ultimately the same.

Riverside was the first of sixteen suburbs designed by Olmsted and Vaux in Massachusetts, New York, Maryland, Illinois, and elsewhere.[62] And while local circumstances differed, they were all designed as sylvan landscapes that would contrast with the harsh rectilinearity of the city itself. So infatuated was Olmsted with his suburban designs that he declared them "the best application of the arts of civilization to which mankind had yet attained."[63] Yet the path from Olmsted's nineteenth century wonderlands to the modern suburbs that sprang up after World War II was not a smooth and easy one. Many suburbs, in fact, were built in the intervening years on a standard grid plan that differed little from the city. It was not until the 1930s, when in the midst of the Depression the Federal Housing Administration began promoting parklike suburbs, that such designs became standard. Modern critics tend to consider these modern developments a much-diminished version of Olmsted's vision of what he called "sylvan beauty." Even so, there is no doubt that the modern suburbs we know today, with their curving streets, dead ends, and cul-de-sacs, are all descended from Olmsted's vision—and, through him, from Central Park.

Olmsted still had one last contribution to make to the American landscape. In 1861, with the outbreak of the Civil War, he took leave of his position as architect in chief of Central Park and, for the next two years served as executive secretary of the US Sanitary Commission—a private organization formed to support the sick and wounded of the Federal Army. Olmsted brought his usual energy and managerial acumen to the task, raising millions of dollars, hiring doctors and nurses, and staffing and equipping field hospitals—even, when necessary, floating ones. But in 1863, the federal government took over the commission's vast operation, and Olmsted was once again at a career impasse. He could have returned to New York, but his frustration with the financial restrictions imposed by the park commissioners was boiling over. In particular, Olmsted chafed against what he called "the systematic small tyranny" of the commission's financial watchdog, Andrew Haswell Green. Working under him, he wrote, "was slow murder," and he was not inclined to put himself in Green's power again.[64]

And so Olmsted headed west. In a surprising turn of events, the visionary designer of Central Park accepted a position as manager of a giant gold mine in California, the Mariposa Estate, just down the Merced River from Yosemite Valley. This time, however, even Olmsted's exceptional managerial talents could not save a failing concern. Within two years, the Mariposa mine was bankrupt, and Olmsted returned east, there to join Vaux in designing Prospect Park. Yet Olmsted's time in California was not wasted. For he not only designed the campus of what would one day become one of the leading universities in the world. He also had a hand in creating the first nature preserve in the United States, which we know today as Yosemite National Park.

Yosemite in the 1860s was just becoming famous among Americans. Thanks to early photographs by Carleton Watkins (1829–1916) and paintings by members of the Rocky Mountain School, the dramatic cliffs and waterfalls of Yosemite Valley became stand-ins for the majestic beauty and raw power of untamed nature.[65] In 1864, in an effort to protect the park from creeping development, President Lincoln's administration ceded the entire region to the State of California, to serve as a pristine natural preserve. To oversee what was in essence the very first state park, California's governor set up a special commission and invited Olmsted to serve as one of its members. Olmsted had already fallen in love with the Yosemite Valley, referring to it in his letters as "sublimely beautiful." He immediately accepted the offer and set to work on what he knew best—a grounds management plan for the new park.[66]

Whether Olmsted should be considered a founder of the national parks movement in the United States is debatable. It can certainly be argued that the honor more properly belongs to personages such as John Muir, who dedicated his life to wilderness preservation, and President Theodore Roosevelt, who oversaw the vast expansion of the system around the turn of the twentieth century.[67] Olmsted's involvement with Yosemite, in contrast, lasted but one year, and in any case, the management of the park by the State of California was not a success. In 1890, the lands of the park reverted back to the federal government, which made Yosemite the second national park in the country (after Yellowstone in 1872). Yet Olmsted's presence at the birth of the national parks movement signifies that it was part of a broader pattern that transformed the American landscape in those years.[68]

Olmsted was the moving spirit behind the creation of urban parks in practically all American cities; he played a major role in the invention and spread of the classic American suburb; and he designed the first college campus in the style of an academic park. All of these are imprinted with his vision of a naturalistic landscape with curvilinear paths and streets and plentiful and varied greenery. All, furthermore, were designed to contrast sharply with the city grid. Olmsted's involvement with the wild national parks illuminates the fact that they expressed much the same vision and served a similar purpose. The urban parks, suburbs, and campuses stand as a contrast and a rebuke to the rectilinear city; the untamed national parks, for their part, do the same to the rural rectilinear grid that encompasses the western United States.

The Scar

And so it is across the American landscape, from great cities to small towns and from cultivated farmland to untouched wilderness. The Great Western Grid covers two-thirds of the land area of the continental United States, creating a uniquely American landscape of unrelenting uniform squares and regularly spaced straight lines, aligned to the points of the compass and stretching from horizon to horizon. The small towns scattered within this landscape recreate the pattern on a miniature scale, and so do American cities, including the very greatest. The old established city of New York was not built as a grid, but its expansion from 1811 onward made its street plan into the world's most famous urban grid. Meanwhile, nearly all American cities founded after independence are made up of arrow-straight streets intersecting one another at right angles. It is indisputably the iconic American urban plan.[69]

And wherever the grid appears in the American landscape, so does its counterpoint—the ungeometrical naturalistic landscape of curvilinear streets that intersect at all angles except ninety degrees. It appears in vast parks at the heart of the gridded city, in urban arcadias such as Central Park in Manhattan, Prospect Park in Brooklyn, or Golden Gate Park in San Francisco. It is there in the belt of modern suburbs, most (though not all) built in the postwar era, with their single-family homes and green yards lined along curving streets and quiet cul-de-sacs, and in the bucolic college campuses large and small sprinkled throughout the United States. It is there in the great national parks, preserved as testaments to the beauty and grandeur of undisturbed nature, as well as in their less famous cousins, the national monuments, state parks, and wildlife refuges. From the famous deep-green forests of Yellowstone to the dramatic reddish-brown cliffs of the Grand Canyon and relatively obscure preserves like the swampy Malheur Natural Wildlife Refuge in Oregon, each one is, in its way, unique. Yet all of them stand in sharp opposition to the endless uniform artificial grid that surrounds them on all sides.

The grid and the anti-grid, two irreconcilable visions of American space, stalk each other across the land. On the one side is the boundless freedom of Jefferson's mathematical rationalism; on the other, the self-reflective caution of Olmsted's anti-mathematical Romanticism. From bustling Manhattan to Big Sky Montana, wherever there is one, the other is surely not far off. Together they forge an uneasy landscape of ideological conflict that defines the United States, the open countryside as well as the busy metropolises. The borderline between them snakes through cities, suburbs, parks, and farmland, like an ideological scar that refuses to heal.

Conclusion: The Chasm

Children of the Grid

On the wintry afternoon of January 2, 2016, several dozen armed men drove up to the offices of the Malheur National Wildlife Refuge in eastern Oregon. Created in 1908 by order of President Theodore Roosevelt, the refuge is known among outdoorsmen and scientists alike for the breathtaking variety of waterfowl and migratory birds that inhabit the Malheur Lake and surrounding marshlands. But the men who arrived at the refuge that Saturday did not come to see the birds. They broke down the doors, entered the empty office buildings, declared the complex their headquarters, and made themselves at home. They constructed a barricade on the only access road to the refuge to prevent anyone from approaching without their knowledge or approval. They set up what they called "defensive positions" around the headquarters and manned a fire lookout tower with an armed guard to watch over the comings and goings at all hours of the day and night. Then they settled down and waited.

So began the occupation of the Malheur National Refuge, which dominated national headlines throughout the United States for nearly a month. By the time it was over, one of the occupation's leaders was dead, and the remaining ones were in police custody, awaiting trial.

While many during and after the events denounced the occupiers for their armed takeover of government property, it was clear that they were not acting for personal gain. Rather, according to their leader, Ammon Bundy of Nevada, they were motivated by a deep sense of grievance against the federal government and the way it was managing western lands. The purpose of the occupation, he explained, was to claim the land of the refuge and hand it over to local ranchers "so that people can reclaim their resources."[1] Once ranchers "can use these lands as free men, then we will have accomplished what we came to accomplish," he said.[2] The land, as far as he was concerned, is a resource for the ranchers and farmers who live on it, and the only role for government was to equitably distribute it among them. For the federal government to restrict ranchers' use of the land to make room for a wildlife refuge, Bundy told an NBC correspondent, "is really a form of tyranny."[3]

Many different roads led to the standoff at the Malheur National Refuge. The occupiers have been tied in reports to extremist groups, such as the "Sovereign Citizen" movement that acknowledges no government authority beyond the county level and believes in the right to armed resistance. Some of the group, including Ammon and his brother Ryan Bundy, have been associated with fringe religious groups, and all of the occupiers and their advocates were supporters of the broad gun rights movement, which rejects government regulation of firearms. All of these factors may have played a role in Ammon Bundy's decision to lead his group to the Malheur National Refuge. Critics of the occupation, who saw the occupiers as dangerous extremists or even domestic terrorists, were quick to point to such unsavory sources of inspiration.[4]

Yet when it came to the occupiers' foremost grievance, the use of western lands, Bundy and his men were drawing on a long-standing American tradition. The notion that the land of America, and in particular the West, exists solely to be used by settlers and farmers goes back to the earliest days of the republic. It was precisely what Jefferson had in mind when he proposed overlaying the American continent with a uniform, all-encompassing, mathematical grid. Erasing both natural features and human history, the grid presented the land as a blank slate, an empty mathematical space ready to be divided up with geometrical precision and used as its new owners saw fit.

When over the course of a century and a half Jefferson's dream was implemented, transforming the western two-thirds of the United States into the graph-paper landscape we know today, it brought in its wake settlers who viewed the land in precisely those terms. To them, the western lands were a limitless but empty expanse, valuable only as a resource for the farmers and ranchers who claim its uniform squares. The armed occupiers who insisted that the sole legitimate use of the lands of the Malheur Wildlife Refuge was as a grazing area for local cattle breeders may have been misguided and violent extremists, as their critics suggest. But they were also giving voice to a deeply held and widely shared belief imprinted over centuries, as government surveyors methodically expanded their precisely measured straight lines and right angles ever westward. Bundy and his men were, in effect, the children of the Great Western Grid.

Indeed, the members of the "Citizens for Constitutional Freedom," as the Malheur occupiers called themselves, had spent their lives on the checkerboard plateaus of the western United States. Some, like LaVoy Finicum, the only person killed in the standoff, were ranchers themselves, but for the most part they were residents of the small, gridded towns that dot the western landscape. Ammon Bundy and his brother Ryan grew up on the family ranch in eastern Nevada, but Ammon later moved to the town of Emmett, Idaho, while Ryan stayed closer to home in the Nevada town of Mesquite. Other members of the group came from similar small towns in Utah, Arizona, Oregon, Washington, and Montana and even as far away as Ohio.[5] Almost without exception, these towns grew up on the right-angle intersections of the arrow-straight roads that mark the boundaries of townships and ranges. Lacking a plan of their own, the towns are aligned with the surrounding fields, their streets running north–south and east–west, creating a miniature urban grid that blends imperceptibly into the surrounding fields.

To be sure, other than in their rectilinear boundaries, where they

The gridded town of Emmett, Idaho, home of Malheur occupation leader Ammon
Bundy, as seen from space. Contains modified Copernicus Sentinel data 2023
processed by Sentinel Hub.

run up against farmers' fields, the federal lands Bundy and his followers were disputing were not themselves gridded. They were, rather,
the regions that were too mountainous and ragged to be neatly carved
into rectangular plots and so were left in government hands when
the surrounding countryside was divided up among the settlers. Encompassed on all sides by the Great Western Grid, the federal grazing
lands of the West are not the grid itself but rather its residue. It is
perhaps inevitable that the battle over land will take place not in the
grid's heartland but on its borderlands, where rectilinear plots give
way to an uncarved landscape. For it is there that the people of the grid
seek to extend its principles to regions that have not yet been brought
under its sway. It is in places like Malheur, within the gridded land but
not of it, that the fight rages on.

For while the occupiers were carrying forth the message of the grid,
their opponents were stalwart devotees of the anti-grid. The scientists, bird-watchers, and nature lovers who frequented the Malheur

preserve did not see it as a blank slate awaiting exploitation. To the contrary, it was to them a dense ecosystem rich in unique animal and plant life, worthy of appreciation, observation, and study. It was humans, to them, who were the unwelcome guests, intruders who must tread carefully so as not to disturb the local wildlife, the refuge's rightful inhabitants. For their efforts, they believed, men and women who take the trouble to immerse themselves in their natural surroundings will be rewarded with a deep appreciation of their place in the world and with spiritual renewal. And so, just as Ammon Bundy and his band were at heart ardent Jeffersonians, the park denizens whom they ousted and displaced were the descendants of Downing, Olmsted, and Muir.

Landscape in Conflict

The standoff at Malheur, for all its drama, was but a single battle in an ongoing war over western lands. In 2014 in Bunkerville, Nevada, rancher Cliven Bundy, father of Ammon and Ryan, confronted federal authorities who had come to seize his cattle. Leading a ragtag army of heavily armed volunteers, he stared down agents of the Federal Bureau of Investigation; the Bureau of Alcohol, Tobacco, Firearms, and Explosives; the Bureau of Land Management; and local police officers until those stood down. In 2016, President Barack Obama created the Bears Ears National Monument on public land in southern Utah. The move was hailed by native tribes, who consider the area sacred ancestral land, but fiercely denounced by local ranchers and businesses as a federal land grab. The critics would later find a friendly ear in President Donald Trump, who in 2017 sliced off more than a million acres from the monument, reducing it by 87 percent. But only four years later, responding to protests from Native American and environmental activists, President Joe Biden restored Bears Ears to its original boundaries. Meanwhile in the Dakotas, at the Standing Rock Indian Reservation, thousands of Native Americans and their allies gathered to protest the construction of the Dakota Access oil pipeline, which was set to traverse tribal lands, as well as the Missouri and Mississippi

Rivers. The demonstrators were cleared out and the pipeline built under the Trump administration, but the battle continues in the courts.[6]

The fight over American land continues, with no end in sight. It runs through the American landscape from great cities to small towns and from cultivated farmlands to untouched wilderness. When oil companies insist on their right to drill where they choose and place pipelines wherever is economically convenient, they are following a long-standing American tradition, but so are the environmentalists and native people who oppose them, fighting for the preservation of pristine natural land. When farmers and ranchers in the West bristle at the restrictions placed on them by the officialdom of the Bureau of Land Management, they see themselves as the heirs of the original settlers, who considered the whole gridded land their patrimony to use and exploit as they wished. Yet much the same can be said of their opponents, who fight to protect the natural landscape from what they consider the alien encroachments of farming, ranching, and industry. They too are heirs to a great American tradition, one that sought to limit and constrain the grid from Central Park to Yosemite Valley.

It is a story as old as the country itself. First came the grid. After the Revolutionary War, as the republic expanded rapidly in territory, population, and wealth, it enthusiastically adopted this simple geometrical scheme. Derived from the rectilinear coordinates of Newtonian absolute space, the grid defined the newly acquired land as devoid of both human history and natural features. Marked like graph paper, with regularly spaced straight lines intersecting at right angles, the rich and varied American landscape became an empty and passive expanse in which enterprising citizens could forge a new nation out of nothing. Whether on the western prairies or the streets of New York, the grid was the bearer of a spirit of independence, opportunity, and optimism. It was the blank slate on which Americans would build their new nation, unconstrained by human history or natural conditions.

Then came the anti-grid. Americans in the middle decades of the nineteenth century had not given up on the grid, the limitless growth and expansion it foretold, or its promise of building a nation free of external constraint. Yet in the crisis of those years, many concluded that this promise also entailed grave risks. The rapid expansion into western territories brought not only conflict with Indians but a struggle between the free and the slave states that culminated in the Civil War. Meanwhile, the gridded artificial cities were settled by wave after wave of immigrants from every corner of the world, bringing their own faiths, customs, and languages, as well as the urban plagues of poverty and crime. The answer, according to Downing, Olmsted, and their collaborators, was to limit the harmful effects of the grid by re-introducing nature into the lives of its inhabitants. Only nature, with its eternal rhythms and a power far beyond any man's, could counter the overweening arrogance of the manmade grid.

The grid and the anti-grid, the abstract mathematical space marked by straight lines and right angles and the naturalistic space of curved lines and irregular angles, are ever present throughout the American landscape. In one America is a blank slate, passively awaiting exploitation by enterprising men who will build their fortunes and their nation. In the other America is an untouched Arcadia, a natural wonderland in which humans must tread lightly. One America is a land of limitless freedom, opportunity, and expansion. The other America, meanwhile, whispers a warning that those who will not follow nature and live solely in a barren mathematical world will lose both the land and themselves. They exist side by side, always at odds yet never apart. Together they create the America we know today.

In the end, the two schemes come down to contrasting ways of perceiving and organizing space. The space of the grid is the space analytic geometry—rational, uniform, and passive, an unresisting arena for unconstrained action. So it is in the massive Jeffersonian grid that covers the western United States, and so it is on the streets of Manhattan and other American cities. The space of the anti-grid is its opposite—irregular and unpredictable, a space in which human action is always constrained and must harmonize with ever-present nature. So it is in Central Park, in Yosemite, and in the ubiquitous suburbs that ring American cities. Both patterns represent long-standing American traditions, the one going back to Thomas Jefferson and Gouverneur Morris, the other to Andrew Jackson Downing and Frederick Law Olmsted. Both bear the imprint of their origins, whether in the explosive growth of the early years of the republic or in the crisis years of the mid-nineteenth century. And both continue to shape Americans' views, beliefs, and even politics to this day.[7]

The political manifestations of the chasm between the two visions play out differently in urban and rural America. On the open plains of the American West, the struggle between the grid and anti-grid aligns neatly with the partisan divide between conservative and liberal America. "We came here and created a blank slate. We birthed a nation from nothing," proclaimed conservative former senator Rick Santorum in 2021, speaking for those who carry on the Jeffersonian vision of an empty gridded land. And indeed, the denizens of the grid—farmers, ranchers, and their small-town neighbors who believe in their absolute right to use the land as they see fit—register as conservative, sometimes of a radical and activist bent. Their nemeses are the liberal environmentalists, native tribes, and government scientists who defend the national parks and wildlife refuges, those isolated natural enclaves in the midst of the graph-paper landscape.

In the great American cities, however, the politics of landscape are almost reversed, and the reasons are not hard to discern. On paper, urban designs like New York's Commissioners' Plan of 1811 seem little different from the Cartesian landscape of the American West, creating a uniform egalitarian blank slate where anything is possible for the enterprising city dweller. But the reality confronted by a New Yorker, Chicagoan, or San Franciscan, whether in the nineteenth century or today, is anything but egalitarian and uniform. For while a rectilinear plot of land in North Dakota is barely distinguishable from one in Nebraska, a city block in New York's Upper East Side is dramatically different—both economically and architecturally—from a city block in East Harlem, just a mile or two away. To poorer urbanites, the conjunction of an egalitarian grid with a starkly stratified reality only highlights the arbitrariness of an unjust social order. To those more economically secure, it points to the precariousness of their own position, never fully legitimate and forever threatened by the teaming masses below. Anxious to escape the tensions brewing in their rectilinear streets, such residents sought refuge in the security of timeless nature, whether out in the wilderness of woods and mountains or at their doorstep in naturalistic city parks. Paradoxically, this made residents of the gray rectilinear streets of American metropolises into dedicated environmentalists and champions of the anti-grid. Meanwhile, the twentieth-century exodus of many upper- and middle-class Americans from the gridded cities to Olmstedian suburbs has made these curvilinear enclaves more politically conservative—and more Republican.

And so the specifics of American politics of space vary from one time and place to another. But through them all runs a long and deep fault line that separates the mathematical rectilinear grid from the naturalistic curvilinear landscape. On the one side are those who believe that America is first and foremost a land of limitless opportunity for its people; on the other are those who fear the consequences of this vision for both the land and its people. It is a fault line that runs from America's great cities to its national parks and from bucolic suburbs to open farmlands. It is a line that separates different races, classes, political allegiances, and deeply held worldviews. It is a line that is ever present across America, and its tensions are manifested in endless political skirmishes over anything from urban zoning to environmental regulations. And on occasion, as at the Malheur National Wildlife Refuge in the winter of 2016, it can erupt in a blaze of violence.

Invisible Patterns

Whether on the streets of a great European capital or on the plains of the western United States, we live in a landscape forged by the power of geometry. The star-shaped plazas of Paris, the converging boulevards of St. Petersburg, and the broad arrow-straight Under den Linden avenue in Berlin are all descended from the Euclidean geometries of Versailles. The rectilinear streets of Manhattan reprise the great grid of the western United States, which is, in turn, a physical incarnation of the Cartesian coordinates of analytic geometry. Geometry even shapes its antithesis: the paths of Central Park, the trails of Yosemite, and even the streets of my own suburban neighborhood were created as a deliberate counterweight to the surrounding geometries. Each of these patterns was shaped by its own history and bestows its own unique meanings on the landscape.

The grand designs of European capitals have their roots in the absolutist monarchy of Louis XIV, founded on the ideals of perfect order and strict hierarchy. The perfect rational order of Euclidean geometry, with its step-by-step hierarchical reasoning and incontestably certain results, embodied both the ideology and power of the absolutist state. In the paths of Versailles as on the avenues of St. Petersburg, the irresistible power of the monarch and the state becomes part and parcel of the universal order of things, as certain and unchallengeable as geometry itself. Even centuries later, with the Bourbons of France and the Romanovs of Russia long gone from the scene, their power to overawe lives on in the geometrically ordered boulevards of their capitals. The kings and emperors may have departed, but the institutions of an all-powerful state are still there, their power inscribed in the incontestable language of Euclidean geometry.[8]

European monarchs were confronted with an ancient land and peoples, and Euclidean geometry proved a powerful tool for recasting their traditions to align with the royal vision. But the leaders of the young American republic, and in particular Jefferson and his allies, saw things very differently. They detested the rigid hierarchies that structured social and political life in the European monarchies, as well as the geometrical constructs that reified and normalized them. In the land of the free, they fervently believed, men and women would start afresh, unencumbered by ancient laws and customs or an oppressive social order. America, to them, was a limitless and empty expanse awaiting its settlers, who through hard work and an enterprising spirit would forge a new nation. And once again geometry offered a way to present their vision and imprint it on the land.

The geometry carved into the American landscape, however, was very different from the traditional Euclidean brand used to buttress the crowned heads of Europe. Analytic geometry, developed over the course of the seventeenth and eighteenth centuries, posited a limitless, empty, and uniform space in which geometrical objects would be defined and created. Unlike Euclidean geometry, it did not impose a single fixed and irrevocable order but offered a boundless space of limitless possibilities. This was precisely what Jefferson and his successors were looking for: they adopted the rectilinear coordinates of analytic geometry and carved them into the landscape of North America. Over the course of a century and a half, an entire continent, rich in natural wonders and human history, was reconfigured as an empty and uniform blank slate, ready for settlement and exploitation.

Two brands of geometry, Euclidean and analytical, were written into the European and American landscapes, inscribing human beliefs on cities and fields. Eventually, perhaps inevitably, they bred their antithesis—a self-conscious and deliberate creation of an anti-geometrical landscape. It first appeared in England, as a retort to both the aesthetics and the politics of Versailles. In response to the strict Euclidean patterns of French formal gardens, English gardeners invented a naturalistic style in which no lines were ever straight and no intersection was ever at right angles. The world of Versailles, in which everyone and everything had their assigned place in a rigid hierarchy, was replaced by a world in which order and hierarchy were hidden and men were left to find their way through the thicket of nature.

Yet ultimately anti-geometry had its greatest impact across the Atlantic, where it became the rival of the all-encompassing Cartesian grid. Whereas in England it aimed to counter the totalitarian regimentation of Euclidean Versailles, in America its purpose was the opposite: to mitigate the perceived chaos engendered by the universal grid. By the middle decades of the nineteenth century, the blank slate that was gridded America came to be viewed by some not as the epitome of freedom and opportunity but as the breeding ground for vice, immorality, and strife. The only way to escape this manmade trap was to reintroduce nature into the grid through urban parks, bucolic suburbs, and pristine natural preserves. As a result, the Cartesian grid and the nongeometrical anti-grid became uncomfortable but inseparable companions in the American landscape. Each encodes a profoundly different understanding of what America is and the ways it relates to its people; both, however, express and reinforce long-standing American traditions.

Euclidean geometry, analytic geometry, and self-conscious anti-geometry shape many—and perhaps most—of the spaces we inhabit every single day. They shape us too. The great boulevards of Paris and St. Petersburg, inspired by the geometries of Versailles, still impress upon us the power of a great centralized state. The Euclidean design of Washington, DC, is carefully calibrated to express the inner tensions and balance of power that sustains the American republic. The gridded prairies of the West still bear Jefferson's vision of a limitless land of unbounded opportunity for its settlers, while the great gridded cities still hum with the energy of men and women making their way in the world and the promise that enterprise and merit will prevail. Meanwhile, winding paths through hills and valleys, across streams, and through green woods still offer a retreat from the world of man to the realm of nature. The patterns are all around us, laden with meaning and history, shaping our attitudes, our beliefs, and our lives.

Remarkably, though, we rarely notice them. Whether Euclidean, Cartesian, or even anti-geometrical, the patterns are well-nigh invisible to us. As we move through them, it is as if they are the only possible reality and could not possibly be different. The great boulevards and star-shaped plazas of European capitals do not appear to us deliberate geometrical constructs in the service of state ideology but simply the way a great city must be. The great grid of the western United States does not seem like the radical mathematical experiment that it is, initiated by the author of the Declaration of Independence. Instead, it simply blends into the western plains, practically invisible, as the only reasonable way to mark the land. And little needs to be said about the parks and suburbs of the anti-grid: these are simply "natural," no matter how much work and resources it takes to design, create, and preserve them.

It is precisely this invisibility that makes these patterns so power-ful. Had they been seen for what they are—artificial human creations encoding worldviews and ideologies—they could be effectively coun-tered. Any ideology, after all, can be resisted, argued against, and ul-timately overthrown. But the geometrical patterns of our world seem to us not fallible human claims but integral parts of the deepest order of the universe. As such, they are unavoidable, irresistible, and ulti-mately invisible. Things simply cannot be otherwise. It is not Louis XIV, or Thomas Jefferson, or Frederick Law Olmsted making those claims; they are simply the order of the world.

And so it has always been with geometry. In 300 BCE, Euclid cre-ated a world that was entirely true and certain, whose order was fixed and its truths unchallengeable. Twenty-three centuries later, he is still at work in our cities and landscapes, making them as necessary and unchallengeable as geometry itself. Euclid's hand is still everywhere, shaping our lives, our beliefs, our politics, our outlooks. We need only open our eyes to see it.

Acknowledgments

Liberty's Grid started its life as one half of a broader project contrasting the underlying geometries of the Old World and the New. But when I showed an early version of the manuscript to my editor at the time, she issued a quick verdict: "Too long!" As a result, what had been a single project became two separate books. *Proof! How the World Became Geometrical* was all about the Euclidean patterns of Versailles and the great capitals of Europe; *Liberty's Grid* is all about the Cartesian patterns of Manhattan and the American West. And so a special acknowledgment goes out to Amanda Moon. This book would never have come to be without you.

My beloved friend and colleague Mary Terrall passed away this year, just as she was beginning to live out her dreams in retirement. I am grateful for her personal and intellectual companionship, and for thirty years of deep and stimulating conversations that have shaped this book. For many discussions, insights, and ideas that (knowingly or unknowingly) found their way into this book, I also thank Ted Porter, Soraya de Chadarevian, Norton Wise, and Robert Frank, my friends in the history of science field at UCLA. Margaret ("Peg") Jacob shared her deep knowledge and insights over many unforgettable lunch dates, and Craig Yirush and Benjamin Madley, my colleagues in the American history field, were generous with their expert advice. Many thanks as well to Avner Ben-Zaken, Raz Chen-Morris, Ofer Gal, Kevin Lambert, Massimo Mazzotti, Jennifer Nelson, Joan Richards, and J. B. Shank for stimulating conversations over many years. Jordan Ellenberg, Michael Harris, Steven Strogatz, and Doron Zeilberger provided mathematical advice, conversation, and support.

I first met Joseph ("Joe") Calamia, my editor at the University of Chicago Press, in 2016, when he reached out to me to discuss possible book projects. It took a few years, but we made it happen. Thank you, Joe, for your support and sage advice, and especially for your patience in seeing the manuscript through all the stages of a complex process. Thanks also to image specialist Matt Lang for help in acquiring and producing the book's images, to production editor Adriana Smith for overseeing the transformation of a manuscript into a book, to Ryan Li for the beautiful cover, to marketing manager Anne Strother for raising *Liberty's Grid*'s public profile, to Angelique Dunn for copyediting the text, and to Carol McGillivray for overseeing the copyediting process. Finally, a heartfelt thanks to the three anonymous reviewers who took the time to read the entire manuscript, and provided insightful, informed, and enormously useful suggestions. *Liberty's Grid* is a much better book because of them.

I am grateful to Dorothy Alexander, Rose Buchanan, George Shaner, and Carmel Wilkes of the National Archives in Washington, DC, for helping me secure an appointment on short notice, track the documents I was looking for, and acquire reproductions of the images I needed. Thanks also to Julie Stoner and Carissa Pastuch of the Geography and Map Division at the Library of Congress for their indispensable help accessing and reproducing nineteenth-century survey maps of the American West.

My longtime agent and friend, Lisa Adams, has been with this project from its earliest beginnings nearly a decade ago and through its numerous incarnations. Her professional acumen, sage advice, and encouragement in challenging times helped me through the ups and downs of a long process.

Daniel Baraz's friendship, loyalty, wisdom, and advice have sustained me for more than half a century now. Although he lives ten time zones away, he joined me on research trips to gardens in the English countryside, Founding Fathers' estates in Virginia, the gridded streets of New York City, and the Euclidean dreamscape of Washington, DC. There is nothing like traveling the world with your closest and oldest friend.

Jutta Sperling, Joshua Feinstein, and Patricia Mazón, my old friends from graduate school, live far away but are always close to my heart. Diane Mizrachi of UCLA, and Peter and Randee Bieler of Malibu are proof positive that it is never too late to enrich one's life with dear and intimate friends. Thank you for many hours of lively conversation about *Liberty's Grid*—and pretty much everything else as well.

My wife, Bonnie, is my true love and my life's companion. Now that our children, Jordan and Ella, are no longer children and have embarked on their lives far away, it is just the two of us again in our little house. I am happy, and grateful, to wake up beside you every morning.

My sisters, Nitza and Michal, have known me all my life. Literally. They stayed in the old country when I moved beyond the ocean, and so we do not see each other as often as we would wish. And yet, I have never felt closer to them than I do today. I love you both and dedicate this book to you.

Notes

Introduction

1. On historical precedents to the American urban grid, see Dan Stanislawski, "The Origin and Spread of the Grid-Pattern Town," *Geographical Review* 36, no. 1 (1946): 105–20, pp. 107–9; Gerrard Koeppel, *City on a Grid: How New York Became New York* (New York: da Capo, 2015); Spiro Kostoff, *The City Shaped: Urban Patterns and Meanings through History* (Boston: Bulfinch, 1991); John W. Reps, *The Making of Urban America: A History of City Planning in the United States* (Princeton, NJ: Princeton University Press, 1965). For more details, see chap. 4 of this book.

2. On the rectilinear division of land in other times and places, see Hildegard Binder Johnson, *Order upon the Land: The U.S. Rectangular Land Survey and the Upper Mississippi Country* (New York: Oxford University Press, 1976), 21–36; Johnson, *The Orderly Landscape: Landscape Tastes and the United States Survey* (Minneapolis: James Ford Bell Library, University of Minnesota, 1977), 9–10; Mahbub Rashid, "The Plan Is the Program: Thomas Jefferson's Plan for the Rectilinear Survey of 1784," paper presented at the ACSA Annual Meeting, Buffalo, NY, October 3–7, 1996, 615–19, p. 617.

3. Johnson, *Orderly Landscape*, 15.

4. An exception could be made for Canada, which adopted the US scheme and extended it with minor modifications into its territory.

5. On the supposed practicality of the grid, see Johnson, *Order upon the Land*; Johnson, *Orderly Landscape*; Norman J. W. Thrower, *Original Survey and Land Subdivision: A Comparative Study of the Forma and Effect of Contrasting Cadastral Surveys* (Chicago: Rand McNally, 1966); William D. Pattison, *Beginnings of the American Rectangular Survey System* (Chicago: University of Chicago Press, 1957); Rashid, "The Plan Is the Program."

6. Early efforts to mark the grid on a small territory in eastern Ohio in the 1780s proved beyond the capabilities of Thomas Hutchins, the most celebrated surveyor of his day and official "Geographer to the United States." On Thomas Hutchins's survey of the Seven Ranges in 1785–86, see chap. 3 of this book.

7. Two and a half centuries after its birth, it is fair to question whether the United States has ever lived up to Jefferson's promise. Many have pointed out that both freedom and opportunity in America have always been strictly curtailed, depending on race and gender. One might endorse Jefferson's vision as an ideal to aspire to or reject it as a mere fig leaf for oppression. Either way, there is no denying the power it has exerted on Americans or the profound ways in which it has shaped the nation's history. This book argues that this power was derived not only from the soaring rhetoric of the Declaration of Independence but also from the down-to-earth ways in which Jefferson's vision reshaped the land.

8. Studies of the western grid are relatively few and far between, and most do not have a historical focus. This is likely due to the fact that scholars tend to view it as practical convenience or a straightforward manifestation of the broader process of modernization, meaningless in itself. The most systematic treatments of the grid can be found in Pattison, *Beginnings of the American Rectangular Land Survey System*; Johnson, *Order upon the Land*. C. Albert White's *History of the Rectangular Survey System* (Washington, DC: US Department of the Interior, 1983) provides an official, and entirely technical, account of the origins and progress of the survey. Norman J. W. Thrower provides an illuminating comparison between the development of gridded and ungridded land over time in his *Original Survey and Land Subdivision*. An exception to the utilitarian view of the grid can be found in the essays of John Brinckerhoff Jackson, and in particular his "Jefferson, Thoreau, and After," in *Landscape in Sight: Looking at America*, ed. Helen Lefkowitz Horowitz (New Haven, CT: Yale University Press, 1997), 175–82.

9. Despite its name, the Cartesian grid was never, in fact, used by Descartes. It was first utilized systematically by Newton, and it came into widespread use by mathematicians in the eighteenth century. It was only in the nineteenth century that the now-familiar grid system came to be named for Descartes, who was honored as the founder of analytic geometry.

10. Most famously, the general theory of relativity implies that space itself bends around massive objects.

11. Descartes's most systematic treatment of extension and matter, his conclusion that they are the same, and the corollary that there can be no empty space can be found in his *Principles of Philosophy*, part 2, articles 11–22. See René Descartes, *The Philosophical Writings of Descartes*, vol. 1, trans. John Cottingham, Robert Smoothoff, and Dugald Murdoch (Cambridge: Cambridge University Press, 1985), 227–32. On the criticism of absolute space by Descartes and his followers, see Alexandre Koyré, *From the Closed World to the Infinite Universe* (Baltimore: Johns Hopkins University Press, 1957), esp. chaps. 5, 7.

12. Newton's most systematic discussion and defense of absolute space can be found in the "Scholium" to the "Definitions" section of book 1 of the *Principia*. His argument against the Cartesian concept of the plenum and its attendant "vortices" can be found in the opening paragraphs of the "General Scholium" to the second edition of the *Principia*. See Florian Cajori, ed., *Sir Isaac Newton's Mathematical Principles of Natural Philosophy and His System of the World*, 2 vols. (Berkeley: University of California Press, 1934), 1:6–12, 2:543–47.

13. For a detailed discussion of the Leibniz-Clarke correspondence, see Koyré, *From the Closed World*, chap. 11. For the text of the correspondence, see H. G. Alexander, ed., *The Leibniz-Clarke Correspondence* (Manchester: Manchester University Press, 1956).

14. On the Declaration of Independence as a Newtonian-inspired text, see I. Bernard Cohen, *Science and the Founding Fathers* (New York: W. W. Norton, 1995), 108–32.

15. Predictably, the colonists in the Carolinas strongly opposed Locke's rectilinear plan, insisting on their right to divide up the land as they saw fit. "If they be not suffered to choose their conveniencyes," they warned, "it may prove a great retarding of a speedy peopling of this country." See *South Carolina Historical Society Collections* (Charleston: South Carolina Historical Society, 1897), 5:284–85.

16. On the role of the transcendentalists, and in particular Henry David Thoreau, in the emergence of the cult of wilderness in the United States, see Roderick Frazier Nash, *Wilderness and the American Mind* (New Haven, CT: Yale University Press, 2014), first published 1967, pp. 84–95. On John Muir as the prophet of the conservation movement, see Nash, 122–40.

Chapter 1

1. The biographical details of Descartes's life are derived from Adrien Baillet, *Vie de M. Descartes* (Paris: Daniel Horthemels, 1691). Descartes's own recollection of his meditations at the inn are found at the beginning of part 2 of the *Discourse on the Method*, in René Descartes, *The Philosophical Writings of Descartes*, trans. John Cottingham, Robert Smoothoff, and Dugald Murdoch (Cambridge: Cambridge University Press, 1985), 1:11–151, 116.

2. Descartes, *Discourse*, 115.

3. Another important development undermining the authority of medieval learning was the rise of humanism, first in Italy and then throughout northern Europe. Beginning in the late fourteenth century, scholars calling themselves humanists launched a devastating critique of "scholastic" knowledge, the knowledge produced and disseminated by the medieval universities. Scholastic knowledge, they claimed, was based on corrupt translations of classical sources and was caught up in meaningless philosophical and theological technicalities. In its place they held up the ideal of pure classical learning, as exemplified in the writings of Cicero and other ancient authors. Though devout Christians themselves, the humanists'

attack on church-authorized learning contributed much to the skeptical attitudes that gained traction in sixteenth-century Europe.

4. Descartes, *Discourse*, 116.

5. For Descartes's account of his course of reasoning, see *Discourse*, part 4, 128–31, and for greater detail, see *Meditations on First Philosophy*, in *The Philosophical Writings of Descartes*, 2:17–62.

6. Descartes's most detailed discussion of matter, space, and the world can be found in *Principles of Philosophy*, part 2, "The Principles of Material Things," in *The Philosophical Writings of Descartes*, 1:223–47. The discussion of the world as mathematical, as well as the last quote, are in Descartes, *Principles of Philosophy*, 2.64, in *The Philosophical Writings of Descartes*, 1:247.

7. No planets beyond Saturn were known until the late eighteenth century.

8. On the mathematical heavens and the nonmathematical Earth of the medieval world, see Michael J. Sauter, *The Spatial Reformation: Euclid between Man, Cosmos, and God* (Philadelphia: University of Pennsylvania Press, 2019).

9. Descartes, *Principles of Philosophy*, 2.64, in *The Philosophical Writings of Descartes*, 1:247.

10. See Giordano Bruno, *On the Infinite Universe and Worlds*, trans. Scott Gosnell (CreateSpace, 2014), first published 1584; Galileo Galilei, "First Day," in *Dialogues concerning Two New Sciences*, ed. Henry Carew and Alfonso de Salvio (New York: Prometheus Books, 1991), first published 1638.

11. In the early twentieth century, Albert Einstein's general theory of relativity overturned some of the key assumptions of the Newtonian world picture. But more than a century after the publication of Einstein's theory in 1915, it can fairly be said that, with the exception of a small number of highly trained mathematical physicists, the vast majority of us still conceive of our world very much as Newton outlined it in his *Principia Mathematica*, published in 1687.

12. The "inverse square law" states that the force of gravity diminishes as the square of the distance from its source. Newton later expanded it into the principle of universal gravitation, which states that any two objects attract each other in direct proportion to their mass and inverse proportion to the square of the distance between them.

13. On Newton's deep familiarity with Descartes and his work, see J. E. McGuire, "Space, Geometrical Objects, and Infinity: Newton and Descartes on Extension," in *Nature Mathematized: Historical and Philosophical Case Studies in Classical Modern Natural Philosophy*, ed. William R. Shea (Dordrecht: D. Reidel, 1983), 69–112. On Newton and Descartes's *La Géométrie*, see Westfall, *Never at Rest*, 88.

14. For Newton's summary of his objections to Descartes's theory of vortices, see the first two paragraphs of his "General Scholium," in *Sir Isaac Newton's Mathematical Principles of Natural Philosophy and His System of the World*, ed. Florian Cajori (Berkeley: University of California Press, 1962), 2:543–47.

15. See Newton, "General Scholium," 546.

16. For more on Descartes's world and its English critics, see Alexandre Koyré, *From the Closed World to the Infinite Universe* (Baltimore: Johns Hopkins University Press, 1957); McGuire, "Space, Geometrical Objects, and Infinity," 69–112.

17. See Isaac Newton, "Scholium" to *Principia*, bk. 1, in *Sir Isaac Newton's Mathematical Principles of Natural Philosophy and His System of the World*, ed. Florian Cajori, vol. 1, *The Motion of Bodies* (Berkeley: University of California Press, 1962), 6.

18. Discussions of Newton's views on space, and his criticism of Descartes, can be found in Koyré, *From the Closed World to the Infinite Universe*, chaps. 7, 10, 11; McGuire, "Space, Geometrical Objects, and Infinity"; Betty Jo Teeter Dobbs and Margaret C. Jacob, *Newton and the Culture of Newtonianism* (Amherst, NY: Humanity Books, 1998), esp. pp. 38–39. Koyré points to the continuity between Newton's views on space and those of his older Cambridge contemporary the Platonist Henry More, who objected to Descartes's identification of space and matter on philosophical grounds. Dobbs and Jacob emphasize Newton's scientific considerations, particularly the fact that the heavenly bodies seem to move through space without drag, from which he deduced that they are moving through empty space.

19. For a discussion of Newton's evolving views on the source of Gravity, see Dobbs and Jacob, *Newton and the Culture of Newtonianism*, 34–56.

20. Newton, "General Scholium," 547.

21. Isaac Newton to Richard Bentley, January 17, 1693, quoted in Koyré, *From the Closed World to the Infinite Universe*, 179.

22. Koyré, *From the Closed World to the Infinite Universe*, 183–84.

23. David Gregory, "Memorandum, December 21, 1705," in I. Bernard Cohen and Richard Westfall, eds., *Newton* (New York: Norton, 1995), 329.

24. The debate between the followers of Descartes and Newton over scientific determinism and God's freedom of action continued in the famous correspondence between Gottfried Wilhelm Leibniz (1646–1716) and Newton's associate the Reverend Samuel Clarke. Leibniz charged that the Newtonian universe, which depended on God's active involvement, was an insult to God's supreme wisdom. Only a flawed creation, he argued, would require repeated divine intervention. Clarke responded that it was precisely God's continued action through the void of absolute space that signified God's omnipresence; the rationally deterministic world favored by Leibniz and the Cartesians was effectively godless. For a detailed account of the debate, see Koyré, *From the Closed World*, chap. 11.

25. Koyré, chap. 11.

26. The full quote is, "This most beautiful system of the sun, planets and comets could only proceed from the counsel and dominion of an intelligent and powerful being." See Newton, "General Scholium" to *Principia*, 2nd ed., in *Sir Isaac Newton's Mathematical Principles*, ed. Florian Cajori, vol. 2, *The System of the World* (Berkeley: University of California Press, 1962), 544.

27. An insightful discussion of the emergence of a Euclidean universe, replacing the medieval Aristotelian cosmos, can be found in Sauter, *Spatial Reformation*.

28. For the complete quote, see Galileo Galilei, "The Assayer," first published 1623, in *Discoveries and Opinions of Galileo*, ed. Stillman Drake (New York: Anchor Books, 1957), 229–80, pp. 237–38.

29. René Descartes, *The Geometry of René Descartes*, trans. and ed. David Eugene Smith and Marcia L. Latham (New York: Dover, 1954).

30. On Descartes's *Geometry*, see Carl B. Boyer, *History of Analytic Geometry* (New York: Dover, 2004), 74–101; Emily Grosholz, "Descartes' *Geometry* and the Classical Tradition," in *Revolution and Continuity: Essays in the History and Philosophy of Early Modern Science*, ed. Peter Barker and Roger Ariew (Washington, DC: Catholic University Press, 1991), 183–196; Judith Grabiner, "Descartes and Problem-Solving," *Mathematics Magazine* 68, no. 2 (April 1995): 83–97. Note that for Descartes, x and y signified not variables but unknown quantities whose values would be determined through algebraic manipulation.

31. Descartes's treatment of this problem can be found in *Geometry*, 335–338. A clear account and assessment of his solution can be found in Grabiner, "Descartes and Problem-Solving," 89–91.

32. "If we say that they (i.e. the curves) are called mechanical because some sort of instrument has to be used to describe them, then we must, to be consistent, reject circles and straight lines, since these cannot be described on paper without the use of compass and a ruler, which may also be termed instruments." Descartes, *Geometry*, 315–16, pp. 40, 43 in the translation.

33. Descartes, 316, p. 40 in the translation. For an example of the instruments Descartes has in mind and his use of them, see Descartes, 319–22, pp. 51–52 of the translation, as well as a clear presentation of these examples in Grabiner, "Descartes and Problem-Solving," 90–91.

34. Other views of space did, however, exist in the ancient world. Atomists, who followed the teachings of Democritus and Epicurus, believed in an infinite void that seems to suggest an absolute space.

35. On Descartes's and Fermat's contrasting approaches to geometry, see Boyer, *Analytic Geometry*, chap. 5, pp. 74–102. On Fermat and his mathematical work, see Michael S. Mahoney, *The Mathematical Career of Pierre de Fermat* (Princeton, NJ: Princeton University Press, 1973). On van Schooten, de Witt, de Lahire, Wallis, and other mathematicians who contributed to the development of analytic geometry in the latter half of the seventeenth century, see Boyer, *Analytic Geometry*, chap. 6, pp. 103–37.

36. See "Newton's Preface to the First Edition" in *Sir Isaac Newton's Mathematical Principles*, ed. Florian Cajori (Berkeley: University of California Press, 1962), 1:xvii.

37. Significantly, Newton chose to present the mathematics of his masterwork, the *Principia*, in Euclidean form, though he pushed the format well beyond what traditional geometers would have recognized. The reasons for this choice are the subject of extensive debate among historians, though the likely reason is that he wanted to limit the criticism of what was already bound to be a controversial work.

Nevertheless, there is no doubt that Newton conceived of the *Principia* in terms of the new analytic geometry and infinitesimal calculus, which reside in absolute mathematical space. Only then did he "translate" it into Euclidean style, which does not assume absolute space. See Niccolò Guicciardini, "Did Newton Use His Calculus in the *Principia*?," *Centaurus* 40 (1998): 303–44; Guicciardini, *Isaac Newton on Mathematical Certainty and Method* (Cambridge, MA: MIT Press, 2009).

38. Isaac Newton, *Enumeratio Linearum Tertii Ordinis*, in *The Mathematical Papers of Isaac Newton*, ed. D. T. Whiteside, vol. 7, *1691–1695* (Cambridge: Cambridge University Press, 1976), 588–655. The text was completed by 1695 and published in 1704 as an appendix to the first edition of Newton's *Opticks* but was largely composed in 1668 or 1670. See Richard S. Westfall, *Never at Rest: A Biography of Isaac Newton* (Cambridge: Cambridge University Press, 1980), 197–202.

39. According to historian of mathematics Carl B. Boyer, in Newton's *Enumeratio* "one finds for the very first time the systematic construction of two axes." Boyer, *Analytic Geometry*, 139.

40. On Locke's admiration for Newton, see Andrew Janiak, "Newton's Philosophy," in *Stanford Encyclopedia of Philosophy*, ed. Edward N. Zalta, published October 13, 2006, last modified July 14, 2021, https://plato.stanford.edu/entries/newton-philosophy/. For his reference to the "incomparable" Newton, see John Locke, *An Essay Concerning Human Understanding*, ed. Kenneth P. Winkler (Indianapolis: Hackett, 1996), 3.

41. Locke, *Essay*, bk. 2, chap. 1.2.

42. Locke, bk. 2, chap. 8.11.

43. For Locke's endorsement of the possibility of empty space, or "a vacuum," see Locke, bk. 2, chap. 13.21–22. The exchange between Locke and Stillingfleet is recounted in Janiak, "Newton's Philosophy." Locke's response to Stillingfleet can be found in Locke, *Essay*, p. 354. The connection between divine and human freedom became a staple of Newtonians' defense of the philosophical and religious underpinnings of Newton's system. As Samuel Clarke and others argued, the determinism of Descartes's and Leibniz's systems deprived both God and man of their freedom. See Steven Shapin, "On Gods and Kings: Natural Philosophy and Politics in the Leibniz-Clarke disputes," *Isis* 72 (June 1981): 187–215, esp. p. 210.

44. All terms are from Newton's *Enumeratio*.

45. See Colin Maclaurin, *A Treatise of Algebra in Three Parts* (London: A. Millar and J. Nourse, 1748), esp. part 3, "Of the Application of Algebra and Geometry to Each Other," pp. 297–366. For Euler's text, see Leonhard Euler, *Introductio in Analysin Infinitorum*, vol. 2 (Lausanne: Marcus-Michael Bousquet, 1748), 329. The diagram, which is referred to in the text, can be found in *Introductio in Analysin Infinitorum*, vol. 1 (Lausanne: Bernuset, Delamolliere, Falque, 1797), fig. 120, plate 14.

46. For more on the politics of Euclidean geometry and its use at Versailles and elsewhere to buttress royal absolutism, see Amir Alexander, *Proof! How the World Became Geometrical* (New York: Farrar, Straus, and Giroux, 2019).

Chapter 2

1. On the gridding of America, its history, and the differences between east and west, north and south, see Ando Linklater, *Measuring America: How the United States Was Shaped by the Greatest Land Sale in History* (New York: HarperCollins, 2002). More technical accounts can be found in William D. Pattison, *Beginnings of the American Rectangular Land Survey System, 1784–1800* (Chicago: University of Chicago Press, 1957); and C. Albert White, *A History of the Rectangular Survey System* (Washington, DC: US Government Printing Office, 1983).

2. A moving and insightful account of Native American attitudes toward their land can be found in Keith H. Basso, *Wisdom Sits in Places: Landscape and Language among the Western Apache* (Albuquerque: University of New Mexico Press, 1996). A broader historical survey of Native American attitudes toward land and land ownership during the colonial era can be found in Allan Greer, *Property and Dispossession: Natives, Empires, and Land in Early Modern North America* (Cambridge: Cambridge University Press, 2018). Greer argues that Indians, much like Europeans, possessed a range of different concepts of landed property, which manifested among different groups in different times and places. Broadly, however, whereas Europeans defined land in terms of its boundaries, native people defined it from central interior locations.

3. On the Catawba map, see Patricia Galloway, "Debriefing Explorers: Amerindian information in the Deslisles' Mapping of the Southeast," in *Cartographic Encounters*, ed. G. Malcolm Lewis (Chicago: University of Chicago Press, 1998), 223–40, pp. 224–26; Alan Taylor, *A Very Short History of Colonial America* (New York: Oxford, 2013), 1–5. The original map is in the Public Records Office, Kew, London, CO 700/6 (1).

4. On the Indian practice of producing maps to communicate with Europeans, see G. Malcolm Lewis, "Indian Maps: Their Place in the History of Plain Cartography," in *Mapping the North American Plains: Essays in the History of Cartography*, ed. Frederick C. Luebke, Frances W. Kaye, and Gary E. Moulton (Norman: University of Oklahoma Press, 1987), 63–80.

5. For Native American artifacts and gestures that approximate the European notion of a "map," see Malcolm Lewis, introduction to Lewis, *Cartographic Encounters*, 1–6; Lewis, "Frontier Encounters in the Field," in Lewis, *Cartographic Encounters*, 9–32; Peter Nabokov, "Orientations from Their Side: Dimensions of Native American Cartographic Discourse," in Lewis, *Cartographic Encounters*, 241–69; Gregory L. Waselkov, "Indian Maps of the Colonial Southeast: Archaeological Implications and Prospects," in Lewis, *Cartographic Encounters*, 205–21. According to Lewis's introduction, no precontact artifacts can reliably be identified as maps.

6. The English captions were almost certainly not in the original but were added by a copyist who transcribed the map on paper.

7. On "fires," see Galloway, "Debriefing Explorers," 224.

8. For maps of Charleston and its gridded street plan in this period, see John W. Reps, *The Making of Urban America: A History of City Planning in the United States* (Princeton, NJ: Princeton University Press, 1965), 175–180.

9. According to anthropologist Peter Nabokov, many American Indian peoples consider the land to be its own best map, as if the topography possessed volitional authority. Before representing it, some native traditions require one to listen to its stories, learn its names, follow it with their feet, and dream in its most propitious locations—and only then represent it on a two-dimensional surface. See Peter Nabokov, "Orientations from Their Side," 243.

10. For the incompatibility of the European and American Indian views of the land, consider the reactions of different tribes to the demand that they submit their territories to surveying and rectilinear demarcations. See Nabokov, "Orientations from Their Side," 244, 248.

11. On the rectilinear surveys and property divisions that preceded the American grid, see Hildegard Binder Johnson, *Order upon the Land: The U.S. Rectangular Land Survey and the Upper Mississippi Country* (New York: Oxford University Press, 1976), 21–36; Johnson, *The Orderly Landscape: Landscape Tastes and the United States Survey* (Minneapolis: James Ford Bell Library, University of Minnesota, 1977), 9–10; Rashid, "The Plan Is the Program," 615–19, p. 617.

12. A few have remarked on the grand ambition of the grid and the staggering effort it took to complete. Hildegard Binder Johnson has called it "mind-boggling" and "the most grandiose and unconditionally rigorous planning design which men ever conceived for dividing their land." Johnson, *Orderly Landscape*, 7, 15. William E. Peters went even further, calling it "one of man's greatest conceptions" and describing the 1785 ordinance that set it in motion as one of "the few important state papers upon which the fundamental rights of mankind are founded." Peters, *Ohio Lands and Their Subdivisions* (Athens, OH: William E. Peters, 1918), 27, 51–52.

13. According to Hildegard Binder Johnson, the nineteenth-century settlers in the upper Midwest wrote a great deal about the land but hardly ever mentioned its most obvious characteristic: it is divided into strict rectilinear plots, which they had arrived to settle and farm. Even today, she notes, it is mostly foreigners who comment on the strangeness of the American system of land division. To the locals, it is all but invisible. See Johnson, *Order upon the Land*, 15–16.

14. On Peter Jefferson, his surveying work, and land speculation, see Linklater, *Measuring America*, chaps. 3, 4; Joel Kovarsky, *The True Geography of Our Country: Jefferson's Cartographic Vision* (Charlottesville: University of Virginia Press, 2014), 10–11. On the hunger for land in the colonies see Linklater, *Measuring America*, chap. 3. For information and images of the Fry-Jefferson Map, see "The Fry-Jefferson Map of Virginia" at https://www.monticello.org/site/house-and-gardens/fry-jefferson-map -virginia, retrieved July 20, 2023. On Thomas Jefferson's recollection of his father and his surveying career, see Thomas Jefferson, "Autobiography," in *Writings*, ed. Merrill D. Peterson (New York: Library of America, 1984), 3–4.

15. On George Washington's land speculations and holdings, see Linklater, *Measuring America*, chaps. 3, 4; Joel Achenbach, *The Grand Idea* (New York: Simon and Schuster, 2002), chaps. 5, 6.

16. As Joel Achenbach puts it, "There was a presumption underlying this forward thrust, a belief that the continental interior was in some fundamental way unoccupied." See Achenbach, *Grand Idea*, 74.

17. See John Winthrop, *Life and Letters*, ed. Robert C. Winthrop (Boston: Ticknor and Fields, 1864), 312.

18. The question of whether and to what extent English colonists in America considered the Indians to be "owners" of the land has generated some historical controversy. Stuart Banner argues that although the original royal charters for the colonies denied that the Indians were owners, in practice the colonists acknowledged their ownership rights by purchasing the land from them. This attitude was reversed, Banner argues, in the early United States, culminating in the 1823 Supreme Court decision in the case of *Johnson v. M'Intosh*, which stated explicitly that while the Indians have "rights of occupancy," they are not owners of the land. Allan Greer complicates the picture by arguing that the colonists brought with them from Europe a range of different notions of "property," none of which corresponded to the modern concept of absolute private property. In this, he argues, they were not so different from the Indians. Consequently, the exchanges between the colonists and the Indians should be viewed as a negotiation between different notions of property rather than as the imposition of a ready-made European scheme on a land that previously had no concept of property. Both Banner and Greer, however, agree that the key factor in the colonists' attitudes was the balance of power between them and the Indians. As the settlers grew more numerous and their military and technological advantage over Native Americans more pronounced, they increasingly ignored and overruled the Indians' claims to land ownership. This was decisively the case in the decades stretching from the aftermath of the Seven Years' War to the Early Republic. See Stuart Banner, *How the Indians Lost Their Land* (Cambridge, MA: Belknap, 2005); Greer, *Property and Dispossession*.

19. See Max Edelson, *The New Map of Empire: How Britain Imagined America before Independence* (Cambridge, MA: Harvard University Press, 2017). Edelson argues that banning a westward expansion was part of an effort by government officials to create an integrated British empire in the Americas, which would be populated primarily by Indians and serve the interests of the mother country.

20. On the contrasting visions of America by the king and his ministers, on the one hand, and the colonists, on the other, see Edelson, *New Map of Empire*. Edelson notes that the king's policies were a direct affront to the colonists' independence and their self-identity as a "settler people."

21. On the companies formed to survey and settle western lands in the middle of the eighteenth century, see Linklater, *Measuring America*, 44–45.

22. See George Washington to William Crawford, September 17, 1767, Founders Online, National Archives, https://founders.archives.gov/documents/Washington/02-08-02-0020.

23. Thomas Jefferson, *A Summary View of the Rights of British America* (Williamsburg, VA: Clementina Rind, 1774). The page numbers that follow refer to the edition

printed in Thomas Jefferson, *Writings*, ed. Merrill D. Peterson (New York: Library of America, 1984), 103–22.

24. Jefferson, *Summary View*, in *Writings*, 118.

25. Jefferson, 119.

26. Jefferson, 119.

27. Jefferson, 105–6.

28. Jefferson, 119.

29. Jefferson, 119–20.

30. On Locke's views on the role of cultivation in land ownership and on the un- productive American Indians, see John Locke, *Second Treatise on Government*, ed. C. B. McPherson (Indianapolis: Hackett, 1980), first published 1690, chap. 5, pp. 18–30. The quote is from chap. 5.37, p. 24. For a discussion and criticism of Locke's view of the Indians as noncultivators, see Banner, *How the Indians Lost their Land*, 46–48.

31. See Thomas Jefferson, *Notes on the State of Virginia* (New York: Penguin Books, 1999), first published in France, 1785, esp. Query VI and Query XI. On Jefferson's attitude toward the Indians and his belief that they must either become "American" or be removed and destroyed, see Anthony F. C. Wallace, *Jefferson and the Indians: The Tragic Fate of the First Americans* (Cambridge, MA: Belknap, 1999).

32. Thomas Jefferson to George Rogers Clark, December 25, 1780, Thomas Jefferson Foundation, https://www.monticello.org/site/jefferson/empire-liberty -quotation. On Jefferson's support for Clark's campaigns, see Joyce Appleby, *Thomas Jefferson* (New York: Times Books, 2003), 103–4.

33. According to *National Geographic*, $15 million in 1803 is equivalent to $342 million in 2020. See Erin Blakemore, "The Louisiana Purchase Was a Bargain. But It Came at a Great Human Cost," *National Geographic*, April 30, 2020, https:// nationalgeographic.com/history/article/louisiana-purchase-bargain-came-great -human-cost.

34. On the Louisiana Purchase, see Appleby, *Thomas Jefferson*, 63–64; Linklater, *Measuring America*, 172–73.

35. Thomas Jefferson to James Monroe, November 24, 1801, in Jefferson, *Writings*, 1096–99, p. 1097.

36. On Jefferson's attitudes toward the Indians, see Wallace, *Jefferson and the Indians*. On his instructions to Lewis and Clark, see Kovarsky, *True Geography*, chap. 4.

37. In his First Annual Message to Congress, dated December 8, 1801, Jefferson approvingly noted the ongoing efforts to "civilize" the Indians, while simultane- ously referring to the "settlement of the extensive country still remaining vacant within our limits." For Jefferson, the presence of the Indians did nothing to change the "fact" that the land was vacant.

38. For Logan's speech, see Jefferson, *Notes on the State of Virginia*, Query VI, pp. 67–68. For a discussion of the story of Logan the Mingo and its significance for Jef- ferson's attitude toward American Indians, see Wallace, *Jefferson and the Indians*, 1–9.

39. As George Washington put it around the time Jefferson was composing the *Notes*, "the gradual extension of our Settlements will as certainly cause the

Savage as the Wolf to retire; both being beasts of prey tho' they differ in shape."
See George Washington to James Duane, September 7, 1783, Founders Online,
National Archives, https://founders.archives.gov/documents/Washington/99-01
-02-11798. In a similar vein, Jefferson referred frequently to the disappearance
of game from the Indians' hunting grounds as a fact of life that must surely ex-
tinguish the native way of life but for which no responsibility could be assigned.
In a letter to Benjamin Hawkins, he noted that "hunting is already insufficient."
Jefferson to Benjamin Hawkins, February 18, 1803, Founders Online, National
Archives, https://founders.archives.gov/documents/Jefferson/01-39-02-0456. In
an address to a delegation of Indian nations on January 10, 1809, he warned them
against the day "when the game shall have left you." Jefferson to Indian nations,
January 10, 1809, Founders Online, National Archives, https://founders.archives
.gov/documents/Jefferson/99-01-02-9516.

40. In the letter to Hawkins as well as in a letter to William Henry Harrison, Jef-
ferson considered the best path for the Indians when "their history will terminate."
Jefferson to William Henry Harrison, February 27, 1803, Founders Online, National
Archives, https://founders.archives.gov/documents/Jefferson/01-39-02-0500.

41. Jefferson to Hawkins, February 18, 1803. Jefferson expressed the same senti-
ment in his address to the delegation of Indian nations on January 10, 1809, telling
them, "In time you will be as we are; you will become one people with us; your blood
will mix with ours; & will spread, with ours, over this great land."

42. Thomas Jefferson, First Annual Message to Congress, December 8, 1801.

43. Jefferson to Hawkins, February 18, 1803.

44. Jefferson to Harrison, February 27, 1803.

45. Jefferson to George Rogers Clark, January 1, 1780 [letter dated in the old
style, January 1, 1779], Founders Online, National Archives, http://founders.archives
.gov/documents/Jefferson/01-03-02-0289.

46. Jefferson to Harrison, February 27, 1803.

47. Jefferson to Henry Dearborn, August 28, 1807, Founders Online, National
Archives, http://founders.archives.gov/documents/Jefferson/99-01-02-6267#. Also
quoted in Appleby, Thomas Jefferson, 108–9.

48. Jefferson to Indian nations, January 10, 1809, Founders Online, National
Archives, https://founders.archives.gov/documents/Jefferson/99-01-02-9516.

49. Jefferson to Dearborn, August 28, 1807. Also quoted in Appleby, Thomas
Jefferson, 108–9.

50. Jefferson suggested settling eastern tribes beyond the Mississippi if necessary, and in this sense he was the originator of the "removal" policy brutally executed some decades later by Presidents Andrew Jackson and Martin Van Buren. Yet
Jefferson considered such settlement only a temporary expediency, not a permanent
solution to the problem of the Indians. Once the east bank of the Mississippi was
densely packed with Europeans, they would, in his view, naturally expand to the
west bank, overrunning the lands where Indians had been settled and driving them
ever farther westward. "When we shall be full on this side," he wrote John Breckin-

ridge on August 12, 1803, "we may lay off a range of states on the western bank . . . & so range after range, advancing compactly as we multiply." See also Wallace, *Jefferson and the Indians*, 225, 275–73.

51. Jefferson, *Notes on the State of Virginia*, Query XIX. See Jefferson, *Writings*, 123–325, pp. 290–91.

52. Jefferson to John Jay, August 23, 1785, in Jefferson, *Writings*, 818–20, p. 819.

53. Jefferson to John Adams, October 28, 1813, in Jefferson, *Writings*, 1304–10, p. 1309.

54. Jefferson to James Monroe, November 24, 1801, in Jefferson, *Writings*, 1096–99, p. 1097.

55. Jefferson to Major John Cartwright, June 5, 1824, in Jefferson, *Writings*, 1490–96, p. 1497.

56. Jefferson to Cartwright, June 5, 1824, p. 1497.

57. For Locke's exchange with Bishop Stillingfleet, see John Locke, *An Essay Concerning Human Understanding*, ed. Kenneth P. Winkler (Indianapolis: Hackett, 1996), 354.

58. Fisher Ames to Christopher Gore, October 3, 1803. See W. B. Allen, ed., *Works of Fisher Ames* (Indianapolis: Liberty Fund, 1983), 2:1462.

59. Jefferson did not, to be sure, argue or believe that America was literally a featureless uniform expanse. In Query VI of the 1785 *Notes on the State of Virginia*, for example, he took on the French naturalist the Comte de Buffon by arguing that New World creatures are at least as large, strong, and vigorous as their Old World counterparts. In Query VI of the *Notes*, he shows his appreciation and even admiration for Native Americans, and in Query XI, he lists the different tribes in the area of Virginia and their territories. But when it came to the settling of the West, the land, for Jefferson, was indeed a vast empty expanse. The rich flora and fauna, forests, and rivers were but resources for the settlers to draw on, or, on occasion (as in the case of deserts and mountain ranges), obstacles to be overcome. The native people must either melt into the mass of settlers or else be removed or destroyed. Neither nature nor people, for Jefferson, occupied the land or could be allowed to stand in the way of the settlement of the vacuous West. See Jefferson, *Notes on the State of Virginia*. On Jefferson as cartographer, see Kovarsky, *True Geography*. On Jefferson's attitude toward the Indians and his belief that they must either become "American" or be removed and destroyed, see Wallace, *Jefferson and the Indians*.

60. On the American West as Newtonian space, see John Brinckerhoff Jackson, "The Order of a Landscape," in *The Interpretation of Ordinary Landscape* (New York: Oxford University Press, 1979), 153–63. On the West as an abstract uniform space free of traditions and unique "places," see John Brinckerhoff Jackson, "A Sense of Place, a Sense of Time," in *A Sense of Place, a Sense of Time* (New Haven, CT: Yale University Press, 1994), 151–62, pp. 154–55.

61. On Jefferson's admiration of Newton, his portrait of Newton and his death mask, and his study of Newtonian science, see I. Bernard Cohen, *Science and the Founding Fathers* (New York: W. W. Norton, 1995), 97–108. On Jefferson's relationship

with William Small, see John Fauvel, "'When I Was Young Mathematics Was the Passion of my Life': Mathematics and Passion in the Life of Thomas Jefferson" (lecture, University of Virginia, Charlottesville, VA, April 15, 1999), https://math.virginia.edu/history/Jefferson/jefferson.htm.

62. On the Declaration of Independence as a Newtonian-inspired text, see Cohen, *Science and the Founding Fathers*, 108–32.

63. Newton's solution was quickly adopted by the broader community of Enlightenment savants. In the words of historian of mathematics Jeremy Gray, "The growing recognition of the merits of Newton's physics cemented a belief that space was three-dimensional, homogenous, isotropic, and to be described as if it was an infinite coordinate grid." Gray, "Epistemology of Geometry," in *Stanford Encyclopedia of Philosophy*, ed. Jeremy Gray and José Ferreirós, published October 14, 2013, last modified July 7, 2021, https://plato.stanford.edu/entries/epistemology-geometry/.

64. Jefferson to William Duane, October 1, 1812, quoted in Fauvel, "When I Was Young," 3.

65. Jefferson to Benjamin Rush, August 17, 1811, quoted in Fauvel, "When I Was Young," 3, and in David Eugene Smith, "Thomas Jefferson and Mathematics," in *The Poetry of Mathematics and Other Essays* (New York: Scripta Mathematica, 1934), 49–70, p. 59.

66. Quoted in Smith, "Jefferson and Mathematics," 59.

67. Fauvel, "When I Was Young," 2–3.

68. Jefferson to Joseph Willard, March 24, 1789, in Jefferson, *Writings*, 947–49, p. 947.

69. Jefferson to Patrick K. Rogers, January 29, 1824, Founders Online, National Archives, https://founders.archives.gov/documents/Jefferson/98-01-02-4015.

70. Thomas Jefferson to Thomas Paine, December 23, 1788, Founders Online, National Archives, http://founders.archives.gov/documents/Jefferson/01-14-02-0156. The letter is mentioned in Fauvel, "When I Was Young," 7.

71. On the Moldboard, see Fauvel, "When I Was Young," 5–6; Cohen, *Science and the Founding Fathers*, 101.

72. Jefferson to William Green Munford, June 18, 1799, in Jefferson, *Writings*, 1063–66, pp. 1063–64.

73. Jefferson to Robert Patterson, March 21, 1811, Founders Online, National Archives, http://founders.archives.gov/documents/Jefferson/03-03-02-0365.

74. Jefferson to Bishop James Madison, December 29, 1811, Founders Online, National Archives, http://founders.archives.gov/documents/Jefferson/03-04-02-0283, quoted in Smith, "Jefferson and Mathematics," 58.

75. E. Millicent Sowerby, ed., *Catalogue of the Library of Thomas Jefferson*, vol. IV (Washington, DC: Library of Congress, 1955).

76. On Jefferson's insistence on high-level mathematical instruction at West Point, see Fauvel, "When I Was Young," 12. On his connection to the École Polytechnique, see Robert Ranquet, "Thomas Jefferson et l'École Polytechnique: à la recherché des chaînons manquant," *Jaune et la Rouge*, no. 536 (June/July 1998).

77. On Jefferson's influence on the mathematical curriculum at the University of Virginia, see *Report of the Commissioner for the University of Virginia*, in Jefferson, *Writings*, 457–73, pp. 462–63; Fauvel, "When I Was Young," 12–13; Smith, "Jefferson and Mathematics," 63–64.

78. On the currency crisis in the early United States, see John Bemelmans Marciano, *Whatever Happened to the Metric System?* (New York: Bloomsbury, 2014), 11–13; Linklater, *Measuring America*, 62–64.

79. On the Morrises' proposal, see Marciano, *Whatever Happened to the Metric System?*; Linklater, *Measuring America*; "Editorial Note: Jefferson's Notes on Coinage," Founders Online, National Archives, accessed June 21, 2022, http://founders .archives.gov/documents/Jefferson/01-07-02-0151-0001.

80. On the "mobs of great cities," see Jefferson, *Notes on the State of Virginia*, Query XIX, in Jefferson, *Writings*, 291. For "Speculating phalanx," see Julian P. Boyd, ed., *Papers of Thomas Jefferson*, vol. 17 (Princeton, NJ: Princeton University Press, 1965), 207.

81. On Robert Morris's hope to establish and run a mint, see "Editorial Note"; Linklater, *Measuring America*, 64.

82. The Spanish origins of the US dollar are preserved in the $ designation—an S bisected by a vertical line.

83. Quoted in Linklater, *Measuring America*, 64.

84. See Linklater, *Measuring America*, 65.

85. Linklater, *Measuring America*, 66.

86. On Jefferson's proposed reform, see Thomas Jefferson, "Plan for Establishing Uniformity in the Coinage, Weights, and Measures of the United States, Communicated to the House of Representatives July 13, 1790," Founders Online, National Archives, https://founders.archives.gov/documents/Jefferson/01-16-02 -0359-0005 and https://founders.archives.gov/documents/Jefferson/01-16-02-0359 -0009. Linklater, *Measuring America*, 107–116; Marciano, *Metric System*, 40–58.

87. The dramatic story of Mechain and Delambre's survey is told in Ken Alder, *The Measure of All Things: The Seven-Year Odyssey and Hidden Error That Transformed the World* (New York: Free Press, 2003).

Chapter 3

1. Quoted in Ando Linklater, *Measuring America: How the United States Was Shaped by the Greatest Land Sale in History* (New York: HarperCollins, 2002), 72; William D. Pattison, *Beginnings of the American Rectangular Survey System* (Chicago: University of Chicago Press, 1957), 130–31.

2. Today the point of beginning is located in the city of East Liverpool, Ohio, where the states of Ohio, Pennsylvania, and West Virginia come together.

3. Hutchins's fondness for one-mile square settlements has led William E. Peters to credit him as the originator of the rectangular survey system and the Great Western Grid. Hutchins's squares, however, were isolated from one another; their

orientation depended on local conditions, not the points of the compass; and consequently they did not combine to form a unified grid. See William E. Peters, *Ohio Lands and their Subdivisions*, (Athens, OH: William E. Peters, 1918), 52–53.

4. On Hutchins and his background, see Linklater, *Measuring America*, 72–74. On the two "Geographers to the United States," see William D. Pattison, *Beginnings of the American Rectangular Land Survey System, 1784–1800* (Chicago: University of Chicago Press, 1957), 69.

5. For the total acreage covered by the rectangular survey in the United States and the states in which it was implemented, see John G. McEntyre, *Land Survey Systems* (New York: John Wiley & Sons, 1978), 31–37; Lola Cazier, *Surveys and Surveyors of the Public Domain, 1785–1975* (Washington, DC: US Government Printing Office, 1976), 108.

6. On Hutchins's survey of the Seven Ranges, see Linklater, *Measuring America*, 71–79; Pattison, *Rectangular Land Survey System*, 105–68; C. Albert White, *A History of the Rectangular Survey System* (Washington, DC: US Department of the Interior, 1983), 17–24; Cazier, *Surveys and Surveyors*, 29–33; Lowell O. Stewart, *Public Land Surveys: History, Instructions, Methods* (Aimes, IA: Collegiate Press, 1935; Minneapolis: Meyers Printing, 1976), 18–19. Citations refer to the Meyers edition.

7. On surveying methods in the eighteenth century, see Linklater, *Measuring America*, 75; Pattison, *Rectangular Land Survey System*, 74–75. For a discussion of the instruments used and their improvement into the nineteenth century, see Francois D. Uzes, *Chaining the Land: A History of Surveying in California* (Sacramento: Landmark Enterprises, 1977), 1–39.

8. On Hutchins's first season of surveying, see Linklater, *Measuring America*, 76–77; Pattison, *Rectangular Land Survey System*, 122–31.

9. For a map of the Seven Ranges, see White, *History of the Rectangular Survey System*, 22, fig. 7.

10. On the survey of the Seven Ranges and the auction of its plots, see Bill Hubbard Jr., *American Boundaries: The Nation, the States, the Rectangular Survey* (Chicago: University of Chicago Press, 2009), 215–32, map on p. 231.

11. Writing in 1910, Amelia Clewly Ford commented on Jefferson's insistence on treating the West as a blank slate to be divided rationally and scientifically. "It is not out of character," she wrote, "to consider Jefferson a doctrinaire reformer, eager to sweep away old irrational methods . . . and on the clean slate of the new west to mark out a plan of boundaries which should advance the cause of science and bring peace to future generations." See Ford, *Colonial Precedents of Our National Land System* (Philadelphia: Porcupine, 1976), 73.

12. Jefferson referred repeatedly in his writings to the United States as an "empire of liberty" or "empire for liberty." See, for example, Thomas Jefferson to George Rogers Clark, December 25, 1780, Founders Online, National Archives, https://founders.archives.gov/documents/Jefferson/01-04-02-0295; Jefferson to the Legislature of the Territory of Indiana, December 28, 1805, Founders Online, National Archives, https://founders.archives.gov/documents/Jefferson/99-01-02-2910; Jeffer-

son to James Madison, April 27, 1809, Founders Online, National Archives, https://
founders.archives.gov/documents/Madison/03-01-02-0163.

13. Linklater, *Measuring America*, 61.

14. On the jumble of claims to western lands by the various states at the end of
the Revolutionary War, see Hubbard, *American Boundaries*, 7–38; Linklater, *Measuring America*, 61; White, *History of the Rectangular Survey System*, 2–10; McEntyre, *Land Survey Systems*, 13–23.

15. As early as the fall of 1780, Congress passed a resolution urging the states to
cede their claims to western lands. See Ford, *Colonial Precedents*, 54.

16. On Jefferson's 1784 report, written in his own hand, see Stewart, *Public Land Surveys*, 2; Cazier, *Surveys and Surveyors*, 15.

17. Thomas Jefferson, "Report on Government for Western Territory," in Thomas
Jefferson, *Writings*, ed. Merrill D. Peterson (New York: Library of America, 1984),
376–78. For an account of Jefferson's proposal, see Hubbard, *American Boundaries*,
106–11. Maps of the states implicit in Jefferson's plan can be found in Hubbard,
American Boundaries, 108; Pattison, *Rectangular Land Survey System*, 18.

18. Jefferson, *Notes on the State of Virginia*, Query XIX, in Jefferson, *Writings*, 123–
325, 290–91.

19. Jefferson, "Report on Government for Western Territory," 376–77.

20. Jefferson, "Report on Government for Western Territory," 377.

21. Thomas Jefferson, "Report of a Committee to Establish a Land Office,"
Founders Online, National Archives, April 30, 1784, http://founders.archives.gov/
documents/Jefferson/01-07-02-0148.

22. On Jefferson's plan for the rectangular survey of the West, see Hubbard,
American Boundaries, 183–85; Pattison, *Rectangular Land Survey System*, 37–67; McEntyre, *Land Survey Systems*, 37–41; Stewart, *Public Land Surveys*, 15–21; D. W. Meinig,
The Shaping of America: A Geographical Perspective on 500 Years of History, vol. 1, (New
Haven, CT: Yale University Press, 1986), 341–42.

23. On Jefferson's definition of the geographical mile and the hundred and his
efforts to introduce the units in Virginia and the United States, see Rashid, "The
Plan Is the Program: Thomas Jefferson's Plan for the Rectilinear Survey of 1784,"
615–19, p. 615; Pattison, *Rectangular Land Survey System*, 44–45; Hubbard, *American Boundaries*, 110–11.

24. The obvious drawback of this system, as Jefferson was well aware, is that
it ignores the curvature of the Earth. Whereas the east–west lines will indeed be
spaced precisely ten geographical miles apart, the intersecting north–south lines,
following the meridians, will inevitably converge as one moves northward. This
means that the northern side of each "square" will be shorter than its southern side,
and the area will not be a true hundred or a true square. Jefferson does not deal with
the problem here, and it remained unresolved until Jared Mansfield took over the
survey in the early 1800s.

25. Jefferson was not the only one to claim credit for the idea of the Great
Western Grid. Hugh Williamson of North Carolina was a member of Jefferson's 1784

committee on western lands and an accomplished mathematician in his own right. In a letter to North Carolina governor Alexander Martin, he notes, "I happen to have suggested the plan to the committee." Williams to Alexander Martin, July 5, 1784, Documenting the American South, University of North Carolina at Chapel Hill, https://docsouth.unc.edu/csr/index.php/document/csr17-0057. Amelia Clewly Ford, however, points out that this is unlikely because the idea of the grid is deeply integrated into Jefferson's plan to reform weights and measures and establish rectangular states. So while we do not have access to the deliberations of the committee, it is safe to conclude that the true originator of the scheme was its chairman, Thomas Jefferson. See Ford, *Colonial Precedents*, 63–67.

26. On Locke's identification of Newtonian space with divine and human freedom, see chap. 1 of this book. On Locke's work for Lord Shaftesbury and the Board of Trade and his proposals for land distribution in South Carolina, see Stuart Banner, *How the Indians Lost Their Land* (Cambridge, MA: Belknap, 2005), 46–48; Ford, *Colonial Precedents*, 19–20; Norman J. W. Thrower, *Original Survey and Land Subdivision* (Chicago: Rand McNally, 1966), 10.

27. On the southern practice of metes and bounds surveying, see Linklater, *Measuring America*, 29.

28. On the New England method of land surveying and appropriation, see White, *History of the Rectangular Survey System*, 8, and Linklater, *Measuring America*, 30. On the competing systems of land distribution in the colonies, roughly divided into the Virginia and New England approaches, see Paul W. Gates, *History of Public Land Law Development* (Washington, DC: US Government Printing Office, 1968), 37–46; Hildegard Binder Johnson, *Order upon the Land: The U.S. Rectangular Land Survey and the Upper Mississippi Country* (New York: Oxford University Press, 1976), 21–25.

29. Jefferson was clearly impressed with the New England method of dividing land into rectilinear plots. Only two years before his death, he recommended dividing Virginia into counties of twenty-four miles square and "wards" of six miles square. See Thomas Jefferson to John Cartwright, June 5, 1824, in Jefferson, *Writings*, 1493, as well as "From Thomas Jefferson to John Cartwright, 5 June 1824," Founders Online, National Archives, https://founders.archives.gov/documents/Jefferson/98-01-02-4313.

30. As late as 1774, the British Privy Council ordered the survey of land before its settlement. However, the council insisted that the survey, unlike Jefferson's later rectangular survey, be carried out "according to the nature and situation of the districts to be surveyed." See Gates, *History of Land Law Development*, 47.

31. On Jefferson's western grid as a utopian landscape for the creation of virtuous citizens see John Brinckerhoff Jackson, "Jefferson, Thoreau, and After," in *Landscape in Sight: Looking at America*, ed. Helen Lefkowitz Horowitz (New Haven, CT: Yale University Press, 1997), 175–82.

32. Consistent with the notion that America was a blank slate, and the settlers all equal, all one-square-mile sections were sold for the same price, regardless of the

quality of the land. This was only changed in 1854. See D. W. Meinig, *The Shaping of America: A Geographical Perspective on 500 Years of History*, vol. 2 (New Haven, CT: Yale University Press, 1993), 241.

33. Thomas Jefferson to Francis Hopkinson, May 3, 1784, Founders Online, National Archives, http://founders.archives.gov/documents/Jefferson/01-07-02-0154.

34. All the quotes are from Jefferson to Hopkinson, May 3, 1784.

35. James Monroe to Thomas Jefferson, April 12, 1785, Founders Online, National Archives, https://founders.archives.gov/documents/Jefferson/01-08-02-0047 .

36. George Washington to William Grayson, August 22, 1785, Founders Online, National Archives, https://founders.archives.gov/documents/Washington/04-03-02-0182.

37. From John C. Fitzpatrick ed., *Journals of the Continental Congress*, vol. 30 (Washington, DC: US Government Printing Office, 1934), 231, May 3, 1786. See also Ford, *Colonial Precedents*, 75; Pattison, *Rectangular Land Survey System*, 89.

38. From the *Journals of the Continental Congress*, vol. 30, p. 423, May 12, 1786. See also Ford, *Colonial Precedents*, 76; Pattison, *Rectangular Land Survey System*, 90.

39. William Grayson to James Madison, May 28, 1786, Founders Online, National Archives, https://founders.archives.gov/documents/Madison/01-09-02-0012.

40. On the Kentucky and Tennessee surveys, see Stewart, *Public Land Surveys*, 3. On the 1790 "metes and bounds" survey of the Virginia Military District in Ohio, see Thrower, *Original Survey and Land Subdivision*, 18.

41. Ford, *Colonial Precedents*, 76.

42. For general discussion of the 1796 debate, see Ford, *Colonial Precedents*, 79–80.

43. *Annals of Congress*, 4th Congress, 1st Session, 329.

44. *Annals of Congress*, 4th Congress, 1st Session, 336.

45. *Annals of Congress*, 4th Congress, 1st Session, 347.

46. Richard Dobbs Spaight to the governor of North Carolina Richard Caswell, June 5, 1785, quoted in Pattison, *Rectangular Land Survey System*, p. 89.

47. "An Ordinance for Ascertaining the Mode of Disposing of Lands in the Western Territory, Passed May 20, 1785," in White, *History of the Rectangular Survey System*, 11–14. The quotes are from pp. 11–12.

48. James Monroe to Thomas Jefferson, January 19, 1786, Founders Online, National Archives, http://founders.archives.gov/documents/Jefferson/01-09-02-0175. Also quoted and discussed in Pattison, *Rectangular Land Survey System*, 31–32.

49. Thomas Jefferson to James Monroe, July 9, 1786, Founders Online, National Archives, http://founders.archives.gov/documents/Jefferson/01-10-02-0045.

50. On the Northwest Ordinance of 1787, see Linklater, *Measuring America*, 181–83; Pattison, *Rectangular Land Survey System*, 16, 33–36; White, *History of the Rectangular Survey System*, 15–16.

51. Pattison, *Rectangular Land Survey System*, 155.

52. Pattison, *Rectangular Land Survey System*, 156–57. According to Linklater, the income from the sales came to $117, 091. Linklater, *Measuring America*, 7.

53. Despite the seeming simplicity of the grid, transferring the rectilinear plots of the West into the hands of farmers proved far more challenging than Jefferson and his allies had envisioned. Confronting these problems led Congress to repeatedly revise the size, price, and terms of the sale of land to settlers throughout the nineteenth century. Between 1789 and 1834 alone, Congress passed 375 different land laws. Even after the system was codified in the Homestead Act of 1862, changes were repeatedly instituted well into the early 1900s. See Richard White, *It's Your Misfortune and None of My Own: A History of the American West* (Norman: University of Oklahoma Press, 1991), 137–54.

54. Rufus Putnam to Fisher Ames, January 6, 1791, in *The Memoirs of Rufus Putnam* (Boston: Houghton Mifflin, 1903), 247. Also quoted in Linklater, *Measuring America*, 119.

55. See "Thomas Jefferson to the Speaker of the House of Representatives," 4 July, 1790, Founders Online, National Archives, https://founders.archives.gov/documents/Jefferson/01-16-02-0359-0005, and "Final State of the Report on Weights and Measures," 4 July 1790, Founders Online, National Archives, https://founders.archives.gov/documents/Jefferson/01-16-02-0359-0009.

56. Quoted in Linklater, *Measuring America*, 123.

57. On the Battle of Fallen Timbers and its impact on the settlement of the West, see Linklater, *Measuring America*, 155–56.

58. Quoted in Linklater, *Measuring America*, 158.

59. "An Act Providing for the Sale of the Lands of the United States, in the Territory Northwest of the River Ohio, and above the Mouth of Kentucky River," May 18, 1796, in *United States Statutes at Large*, vol. 1, 4th Congress, 1st Session, chap. 29, https://en.wikisource.org/wiki/United_States_Statutes_at_Large/Volume_1/4th_Congress/1st_Session/Chapter_29. For a discussion of the 1796 act, see White, *History of the Rectangular Survey*, 21–22; Linklater, *Measuring America*, 156.

60. On the strains on Putnam's survey, see Linklater, *Measuring America*, 162–67.

61. On Jefferson's struggle to appoint Republicans to replace Federalist officeholders left over from the Adams administration, see Joyce Appleby, *Thomas Jefferson* (New York: Times Books, 2003), 36–40.

62. Albert Gallatin to Rufus Putnam, September 21, 1803, in *The Memoirs of Rufus Putnam and Certain Official Papers and Correspondence*, ed. Rowena Buell (Boston: Houghton, Mifflin, 1903), 439.

63. Rufus Putnam to Albert Gallatin, February 15, 1804, in Buell, *Memoirs of Rufus Putnam*, 440–41.

64. Buell, *Memoirs of Rufus Putnam*, 125.

65. Jared Mansfield, *Essays, Mathematical and Physical: Containing New Theories and Illustrations of Some Very Important and Difficult Subjects of the Sciences* (New Haven, CT: W. Morse, 1801). The biographical information is from Joe Albree, "Jared Mansfield (1759–1830): Janus Figure in American Mathematics," in *History of Undergradu-*

ate Mathematics in America: Proceedings of a Conference Held at the United States Military Academy, West Point, New York, June 21–24, 2001, ed. Amy Shell-Gellasch (West Point, NY: Department of Mathematical Sciences, 2002), 73–94.

66. Quoted in Linklater, *Measuring America*, 175.

67. For Mansfield's instructions to his deputy surveyors, see "General Instructions to Deputy Surveyors, Jared Mansfield, 1804," in White, *History of the Rectangular Survey System*, 237–38.

68. On Mansfield's gridding of the West using principal meridians, base lines, townships, and ranges, see Linklater, *Measuring America*, 175–77; Hubbard, *American Boundaries*, 271–74; Cazier, *Surveys and Surveyors*, 17.

69. Today there are thirty-one principal meridians and their accompanying base lines in the lower forty-eight states, the first six of them numbered, the others named. Five additional principal meridians and base lines are located in Alaska. See Cazier, *Surveys and Surveyors of the Public Domain*, 17.

70. On the correction lines, including "standard parallels" and "guide meridians," see Cazier, *Surveys and Surveyors of the Public Domain*, 17; Ronald E. Grim, "Mapping Kansas and Nebraska: The Role of the General Land Office," in *Mapping the North American Plains: Essays in the History of Cartography*, ed. Frederick C. Luebke, Frances W. Kaye, and Gary E. Moulton (Norman: University of Oklahoma Press, 1987), 127–44, p. 128. Surveying practices were standardized in 1855 with the publication by the General Land Office of "Instructions to the Surveyors General of Public Lands of the United States," reprinted in White, *History of the Rectangular Survey System*, 457–500.

71. On the changing minimal size of land plots for sale by the federal government, see Linklater, *Measuring America*, 180–81; Stewart, *Public Land Surveys*, 12–24; Meinig, *Shaping of America*, 2:242.

72. Tecumseh's missive to Jefferson is quoted in Allan Greer, *Property and Dispossession: Natives, Empires, and Land in Early Modern North America* (Cambridge: Cambridge University Press, 2018), 310.

73. See Thomas Jefferson to Benjamin Hawkins, February 18, 1803, Founders Online, National Archives, https://founders.archives.gov/documents/Jefferson/01-39-02-0456; Jefferson to William Henry Harrison, February 27, 1803, Founders Online, National Archives, https://founders.archives.gov/documents/Jefferson/01-39-02-0500; Jefferson to John Breckinridge, August 12, 1803, Founders Online, National Archives, https://founders.archives.gov/documents/Jefferson/01-41-02-0139.

74. Even without grounding in official policy, Jefferson managed to acquire two hundred thousand square miles of territory from the Indians and open it up for survey and settlement. This number does not include the vast territories of the Louisiana Purchase. See Anthony F. C. Wallace, *Jefferson and the Indians: The Tragic Fate of the First Americans* (Cambridge, MA: Belknap, 1999), 239–40.

75. On Monroe's proposal for the removal of the Indians, see Meinig, *Shaping*

of America, 2:81–83. On the 1823 Supreme Court decision in the case of *Johnson v. M'Intosh*, see Stuart Banner, *How the Indians Lost Their Land* (Cambridge, MA: Belknap, 2005), 11, 178–90.

76. On the Indian Removal Act, see Meinig, *Shaping of America*, 2:88–89; William E. Unrau, *The Rise and Fall of Indian Country* (Lawrence: University of Kansas Press, 2007), 9.

77. On the Black Hawk War and the Trail of Tears, see Meinig, *Shaping of America*, 2:83–91.

78. See Indian Removal Act of 1830, Section 3. On the 1834 Indian Trade and Intercourse Act, see Unrau, *Rise and Fall of Indian Country*, 1–9.

79. Thomas Jefferson to John Breckinridge, August 12, 1803, Founders Online, National Archives, https://founders.archives.gov/documents/Jefferson/01-41 -02-0139.

80. On the law of 1853, see Unrau, *Rise and Fall of Indian Country*, 9. For the end of Indian Country, see Unrau, 141–49.

81. Hugh Williamson to Alexander Martin, July 5, 1784, Documenting the American South, University of North Carolina at Chapel Hill, https://docsouth.unc .edu/csr/index.php/document/csr17-0057.

82. William Grayson to George Washington, April 15, 1785, Founders Online, National Archives, https://founders.archives.gov/documents/Washington/04-02-02 -0359.

83. See Johnson, *Order upon the Land*, 20 and throughout; Thrower, *Original Survey and Land Subdivision*, 67, and Ted Steinberg, *Down to Earth: Nature's Role in American History* (Oxford: Oxford University Press, 2002).

84. As early as 1670, settlers in South Carolina protesting against the Earl of Shaftesbury's and John Locke's rectilinear land division plan warned that "it may prove a great retarding of a speedy settlement of the country." See *Collections of the South Carolina Historical Society*, vol. V (Charleston, SC: South Carolina Historical Society, 1897), 284–85, quoted in Ford, *Colonial Precedents*, 343.

85. John Wesley Powell (1834–1902), celebrated western explorer and second director of the US Geological Survey, severely criticized the grid as unsuitable for the settlement of the West. The division of land, he argued in his 1878 *Report on the Lands of the Colorado*, should be adjusted to local conditions and in particular to the presence or absence of water. See Johnson, *Order upon the Land*, 18–19; Michael Bryson, *Visions of the Land: Science, Literature, and the American Environment from the Era of Exploration to the Age of Ecology* (Charlottesville: University Press of Virginia, 2002), 80–101.

86. On the impracticalities of the grid in the Midwest, see Johnson, *Order upon the Land*, esp. chap. 6, pp. 116–50; Johnson, *The Orderly Landscape: Landscape Tastes and the United States Survey* (Minneapolis: James Ford Bell Library, University of Minnesota, 1977); Thrower, *Original Survey and Land Subdivision*.

87. DeWitt Clinton, mayor of New York City and governor of New York in the

early nineteenth century, argued that straight roads are bad because "straightness means up and down." Quoted in Johnson, *Order upon the Land*, 167.

88. On the roads of the grid, see Johnson, *Order upon the Land*, 167–68, 170–71; Thrower, *Original Survey and Land Subdivision*, 86–117. On numerous and expensive angled bridges, see Thrower, *Original Survey and Land Subdivision*, 98.

89. On the settlers purchasing their tracts in local land offices rather than in the cities they set out from, see Johnson, *Order upon the Land*, 127–140. The local land offices were established by the General Land Office, a federal agency founded in 1812 to facilitate the transfer of public lands to settlers. See Cazier, *Surveys and Surveyors*, 60.

90. Note that as early as 1786, Rufus King warned that the requirements of the rectilinear survey "will greatly delay the survey of the said territory." See *Journals of the Continental Congress*, vol. 30, p. 423, May 12, 1786.

91. A telling example of the power of the grid to shape the attitudes of its residents is the seemingly irrational attachment of midwestern farmers to the impractical and inconvenient rectilinear pattern of their roads. As early as 1847, William Mitchell Gillespie complained about "the unwillingness of farmers to allow a road through their farms in a winding line" because "they attach more importance to the squareness of their fields than to the improvement of the lines of their roads." The attitude remained so entrenched that in 1935 the Iowa General Assembly prohibited the construction of diagonal roads around Des Moines. In 1963, the law was finally repealed to allow for the construction of the interstate highway system, but only in the face of strong opposition, which resurfaced every time new nonrectilinear highways were proposed. See Johnson, *Order upon the Land*, 167, 171; Johnson, *Orderly Landscape*, 14.

92. For a full discussion of the gardens of Versailles, their geometrical structure, and their political and ideological power, see Amir Alexander, *Proof! How the World Became Geometrical* (New York: Farrar, Straus, and Giroux, 2019), 131–83. See also the discussion of Versailles's influence on the design of Washington, DC, in chap. 4 of this book.

93. "The lands are of so versatile a nature," Washington wrote, "that to the end of time they will not be . . . purchased either in townships or by square miles." George Washington to William Grayson, August 22, 1785, quoted in Pattison, *Rectangular Land Survey System*, 89.

94. For a discussion of Jefferson's instructions to Lewis and Clark, see Andrea Wulf, *Founding Gardeners: The Revolutionary Generation, Nature, and the Shaping of the American Nation* (New York: Vintage Books, 2011), 156–59.

95. Quoted in Andro Linklater, *Measuring America: How the United States Was Shaped by the Greatest Land Sale in History* (New York: Plume, 2002), 78.

96. Quoted in Wade Graham, *American Eden* (New York: Harper Perennial, 2009), 62.

97. The most influential thesis on the historical formation of an "American

336 NOTES TO PAGES 163–167

character" is Frederick Jackson Turner's "frontier thesis." The American "national character," Turner argued, was forged in the encounter with wilderness on the western frontier. While dominant in the early twentieth century, Turner's thesis came under withering criticism in later years. Frederick Jackson Turner, "The Significance of the Frontier in American History," in *The Frontier in American History* (New York: Barnes and Noble, 2009), 1–24. The essay was based on Turner's talk at the American Historical Association meeting in 1893.

Chapter 4

1. *Athenaeum*, June 16, 1849, 629.
2. Manahatta, the Lenape name for the island, is generally translated as "Island of Hills." See, for example, Gerrard Koeppel, *City on a Grid: How New York Became New York* (New York: da Capo, 2015), xvi.
3. For the 1811 Commissioners' Plan for Manhattan, see later in this chapter, as well as Hillary Ballon, ed., *The Greatest Grid: The Master Plan of Manhattan, 1811–2011* (New York: Museum of the City of New York and Columbia University Press, 2012). For the expansion of the grid around the middle of the nineteenth century, see Matthew Dripps, "Map of the City of New York Extending Northward to Fiftieth Street," issued 1851, in *Manhattan in Maps*, ed. Paul E. Cohen and Robert T. Augustyn (New York: Dover, 1997), 110–11. For the growing population of New York, see Howard P. Chudacoff and Judith E. Smith, *The Evolution of American Urban Society* (Upper Saddle River, NJ: Prentice Hall, 2000), 7, 55; Ballon, *Greatest Grid*, 27.
4. On Manhattan as a Cartesian grid, see Reuben Rose Redwood, "Rationalizing the Landscape: Superimposing the Grid upon the Island of Manhattan" (master's thesis, Pennsylvania State University Department of Geography, 2002).
5. Thomas Jefferson to William Short, September 8, 1823, in *The Writings of Thomas Jefferson*, ed. H. A. Washington (New York: Riker, Thorne, 1854), 8:309–10, https://founders.archives.gov/documents/Jefferson/98-01-02-3750.
6. Thomas Jefferson, *Notes on the State of Virginia*, Query XIX, in Jefferson, *Writings*, ed. Merrill D. Peterson (New York: Library of America, 1984), 123–325, pp. 290–91.
7. For an overview of Jefferson's role in the founding of the federal capital, see C. M. Harris, "Washington's Gamble, L'Enfant's Dream: Politics, Design, and the Founding of the National Capital," *William and Mary Quarterly* 56, no. 3 (July 1999): 527–64.
8. For a detailed account of Jefferson, L'Enfant, and their fight over the proper geometry for the federal capital, see Amir Alexander, "The Euclidean Republic," chap. 7, in *Proof! How the World Became Geometrical* (New York: Farrar, Straus, and Giroux, 2019), 229–57. See also Elizabeth S. Kite, *L'Enfant and Washington 1791–1792* (Baltimore: Johns Hopkins University Press, 1929); John W. Reps, *The Making of Urban America* (Princeton, NJ: Princeton University Press, 1965), 240–56; Spiro Kostof,

The City Shaped: Urban Patterns and Meanings through History (Boston: Bulfinch, 1991), 209–11.

9. See Pierre Charles L'Enfant, Memorandum presented to Washington, March 26, 1791, Founders Online, National Archives, https://founders.archives.gov/documents/Washington/05-08-02-0005. See also Kite, *L'Enfant and Washington*, 47. In his correspondence, L'Enfant regularly refers to the United States as an empire. See, for example, his letter to Washington of September 11, 1789, in which he offered his services as chief designer. L'Enfant to George Washington, September 11, 1789, Founders Online, National Archives, https://founders.archives.gov/documents/Washington/05-04-02-0010.

10. On European monarchs' rush to create their own "Versailles," see Alexander, *Proof!*, 187–95.

11. Jefferson's proposal for the width of the streets is included in his Memorandum, August 29, 1790, Founders Online, National Archives, https://founders.archives.gov/documents/Washington/05-06-02-0176. Jefferson's estimates for the size of the city, as well as his earlier diagram for the layout of the city streets, is in his Memorandum, September 14, 1790, Founders Online, National Archives, https://founders.archives.gov/documents/Washington/05-06-02-0209.

12. For the size of Philadelphia in 1790, see Reps, *Urban America*, 246.

13. See L'Enfant, Memorandum presented to Washington, March 26, 1791, Founders Online, National Archives, https://founders.archives.gov/documents/Washington/05-08-02-0005, quoted in Kite, *L'Enfant and Washington*, 47.

14. Quoted in J. J. Jusserand, "Major L'Enfant and the Federal City," introduction to Kite, *L'Enfant and Washington*, 24.

15. On the rise and fall of L'Enfant, see Reps, *Urban America*, 240–56. On his relationship with Jefferson and Jefferson's role in his fall, see Harris, "Washington's Gamble," esp. 544–50.

16. The story of Washington's plan to be entombed under the Capitol dome and his motives for doing so are told in C. M. Harris, "Washington's Gamble."

17. For Jefferson's advocacy of the "chequer board" plan for cities and towns, including New Orleans, see Thomas Jefferson to Constantin François Chasseboeuf Volney, February 8, 1805, Founders Online, https://founders.archives.gov/documents/Jefferson/99-01-02-1123, discussed and quoted in Reps, *Urban America*, 317.

18. On Jefferson's correspondence with Harrison on Jeffersonville, see William Henry Harrison to Thomas Jefferson, August 6, 1802, Founders Online, National Archives, https://founders.archives.gov/documents/Jefferson/01-38-02-0156; Jefferson to Harrison, February 27, 1803, Founders Online, National Archives, https://founders.archives.gov/documents/Jefferson/01-39-02-0500, discussed and quoted in Reps, *Urban America*, 317–19.

19. On the division of townships into sections and on the segments set aside for public use, see William D. Pattison, *Beginnings of the American Rectangular Land Survey System, 1784–1800* (Chicago: University of Chicago Press, 1957), 92–96; C. Albert

White, *History of the Rectangular Survey System* (Washington, DC: US Department of the Interior, 1983), 11–15, 20, fig. 5.

20. On the emergence of new cities in the American West, see D. W. Meinig, *The Shaping of America*, vol. 2, *Continental America* (New Haven, CT: Yale University Press, 1993), 352–74. On the urban grid of western cities, see Meinig, 384–96.

21. On western towns developing along the grid, see Reps, *Urban America*, 216–217, as well as discussions of specific western cities throughout the volume. On western towns as an extension of the grid, see also John Brinckerhoff Jackson, "Jefferson, Thoreau, and After," in *Landscape in Sight* (New Haven, CT: Yale University Press, 1997), 175–82, esp. 178.

22. On Mohenjo-Daro and Harappa, see Dan Stanislawski, "The Origin and Spread of the Grid-Pattern Town," *Geographical Review* 36, no. 1 (1946): 105–20, pp. 107–9; Koeppel, *City on a Grid*, 1–2; Spiro Kostoff, *The City Shaped: Urban Patterns and Meanings through History* (Boston: Bulfinch, 1991), 103–4.

23. On the ground plans of Babylon and Sargon's capital of Dur-Sarginu (or Dur-Sharrukin), see Stanislawski, "Grid-Pattern Town," 110; Kostoff, *City Shaped*, 104.

24. On the Roman grid plan, see Stanislawski, "Grid-Pattern Town," 116–17; Kostoff, *City Shaped*, 107–8.

25. On medieval bastides and grid cities, see Stanislawski, "Grid-Pattern Town," 118–19; Kostoff, *City Shaped*, 108–11.

26. From "The Laws of the Indies," quoted in Reps, *Urban America*, 29.

27. On early Spanish settlements in America and Laws of the Indies towns, see Reps, *Urban America*, 26–32; Kostoff, *City Shaped*, 111–16; Axel I. Mundigo and Dora P. Crouch, "The City Planning Ordinances of the Laws of the Indies Revisited: Part I; Their Philosophy and Implications," *Town Planning Review* 48, no. 3 (July 1977): 247–68; Dan Stanislawski, "Early Spanish Town Planning in the New World," *Geographical Review* 37, no. 1 (January 1947): 94–105.

28. For the plan of Mendoza, see Angel Rama, *The Lettered City* (Durham, NC: Duke University Press, 1996), opposite p. 1. For the plan of San Fernando de Béxar, see Reps, *Urban America*, 37, fig. 17. For the original layout of Los Angeles, see Mundigo and Crouch, "City Planning Ordinances," 397–418.

29. On the parallels between Vitruvius and the Laws of the Indies, see Stanislawski, "Early Spanish Town Planning in the New World."

30. Quoted in Reps, *Urban America*, 30.

31. Holme's description of the city plan for Philadelphia is quoted in Reps, *Urban America*, 161.

32. For a different interpretation of the contrast between traditional grid towns and New York, see Peter Marcuse, "The Grid as City Plan: New York City and Laissez-Faire Planning in the Nineteenth Century," *Planning Perspectives* 2 (1987): 287–310. Marcuse distinguishes between traditional "closed grid" cities and the "open grid" city of New York, identifying the former with a precapitalist economy and the latter with laissez-faire capitalism.

Chapter 5

1. For a reconstruction of what Manhattan may have looked like in 1609, see Robert T. Augustyn and Paul E. Cohen, eds., *Manhattan in Maps, 1527–1024* (New York: Dover, 2014), xxii–xxiii. On the death of John Colman, Hudson's second mate, see "New York's Coldest Case: A Murder 400 Years Old," *New York Times*, September 4, 2009, http://www.nytimes.com/2009/09/05/nyregion/05murder.html.

2. It was only in the early twentieth century that Norwegian explorer Roald Amundsen managed to sail from the Atlantic to the Pacific Ocean through the Canadian Arctic. It took him from 1903 to 1906 to complete the passage, and it was of no commercial value.

3. *Bever* is Dutch for "beaver."

4. On the population of New Amsterdam, see Gerrard Koeppel, *City on a Grid: How New York Became New York* (New York: da Capo, 2015), 7.

5. Quoted in Koeppel, *City on a Grid*, 9.

6. Quoted in Koeppel, *City on a Grid*, 7.

7. For the painting *Five Points, 1827*, see Hilary Ballon, ed., *The Greatest Grid: The Master Plan of Manhattan, 1811–2011* (New York: Museum of the City of New York and Columbia University Press, 2012), 21, fig. 4. Ballon attributes the painting to George Catlin.

8. On Goerck's surveys, see Koeppel, *City on a Grid*, 20–22, 25–27.

9. Joseph François Mangin to Alexander Hamilton, January 11, 1799, quoted in Koeppel, *City on a Grid*, 31.

10. Joseph François Mangin to Common Council, December 17, 1799, quoted in Koeppel, *City on a Grid*, 37.

11. On Mangin's map of New York, including a reproduction, see Cohen and Augustyn, *Manhattan in Maps*, 82–85.

12. On the fate of Mangin's plan, in greater detail, see Koeppel, *City on a Grid*, 49–63. The quote is on p. 55.

13. Quoted in Koeppel, *City on a Grid*, 56.

14. The text of the act can be found in Ballon, *Greatest Grid*, 30–31.

15. Ballon, 30.

16. Ballon, 30.

17. On Gouverneur Morris's complaints about Rutherfurd, see Koeppel, *City on a Grid*, 78.

18. George Washington to Thomas Jefferson, March 4, 1784, Founders Online, National Archives, http://founders.archives.gov/documents/Jefferson/01-07-02-0006.

19. On Gouverneur Morris's character, affairs, and career, see Richard Brookhiser, *Gentleman Revolutionary: Gouverneur Morris, the Rake Who Wrote the Constitution* (New York: Free Press, 2003). For Morris's loss of his leg and the attendant rumors, see Brookhiser, 61–65. John Jay's letter is quoted in Brookhiser, 61, as well as in Koeppel, *City on a Grid*, 82.

20. Quoted in Brookhiser, *Gentleman Revolutionary*, 84.

21. Brookhiser, 84.

22. Brookhiser, 84–85.

23. Koeppel, *City on a Grid*, 83.

24. Morris's diary entry for January 10, 1807, is quoted in Koeppel, *City on a Grid*, 87.

25. The scarcity of solid evidence has left much room for speculation on the commission's work. For two recent and quite different accounts, see Koeppel, *City on a Grid*, esp. 89–115; and Marguerite Holloway, *The Measure of Manhattan: The Tumultuous Career and Surprising Legacy of John Randel Jr., Cartographer, Surveyor, Inventor* (New York: W. W. Norton, 2013), esp. 54–66. Koeppel contends that the grid pattern was arrived at almost by chance in the final months of the commission's term, as the easiest plan to map with time running out. Holloway argues (more plausibly, in my opinion) that the grid plan had guided Randel's surveying work for several years before it was presented to the assembly.

26. Gouverneur Morris to Simeon De Witt, October 2, 1807, quoted in Koeppel, *City on a Grid*, 91.

27. The account of Morris's dissatisfaction with Loss is from his letter to Simeon De Witt, October 2, 1807. See Koeppel, *City on a Grid*, 91–92; Holloway, *Measure of Manhattan*, 54–55.

28. On Randel's early work in Manhattan, see Koeppel, *City on a Grid*, 100–101.

29. On Randel's extension of Goerck's Common Lands grid, see Koeppel, *City on a Grid*, 101–2; Holloway, *Measure of Manhattan*, 55.

30. On Randel's innovative instruments, see Holloway, 66–88.

31. On the challenge to surveyors of magnetic variation, in Manhattan and elsewhere, and on Randel's efforts to overcome it, see Holloway, 67–74.

32. Martha Lamb, *The History of the City of New York* (New York: A. S. Barnes, 1877), 3:572, quoted in Koeppel, *City on a Grid*, 103, and Holloway, *Measure of Manhattan*, 60.

33. Holloway, 60.

34. See Koeppel, *City on a Grid*, 134; Holloway, *Measure of Manhattan*, 88.

35. John Randel, "City of New York," *Valentine's Manual*, 1864, 848.

36. On Mills's suit against Randel, see Koeppel, *City on a Grid*, 102; Holloway, *Measure of Manhattan*, 61. Details of the suit can be found in John Mills v. John Randel, 1810R-47, Division of Old Records, New York County Clerk's Office.

37. Legal challenges to Randel's survey, and later to the implementation of the grid, were filed repeatedly by property owners and dragged on for decades. All proved fruitless. See Koeppel, *City on a Grid*, 138.

38. On De Witt's complaint to city authorities and the new state act, see Koeppel, *City on a Grid*, 105–6; Holloway, *Measure of Manhattan*, 61–62. The act is in *Laws of the State of New York*, 1809, chap. 103 (March 24, 1809).

39. Holloway, *Measure of Manhattan*, 62.

40. Quoted in Koeppel, *City on a Grid*, 110–12.

41. On Randel's work in Manhattan from 1811 to 1821 and on his maps, see Koeppel, *City on a Grid*, 111–12; Holloway, *Measure of Manhattan*, 102–21; Cohen and Augustyn, *Manhattan in Maps*, 86–97; Ballon, *Greatest Grid*, 34–36, 42–43, 68–70.

42. On the contents of the map, see Ballon, *Greatest Grid*, 33–34.

43. "Remarks of the Commissioners of the 1811 Plan," in Ballon, *Greatest Grid*, 40–42. Quote is from p. 40.

44. See, for example, Françoise Choay, *The Modern City: Planning in the 19th Century* (New York: George Braziller, 1969), 15–22.

45. "Remarks of the Commissioners of the 1811 Plan," in Ballon, *Greatest Grid*, 40.

46. Clement Clarke Moore ["A Landholder," pseud.], *A Plain Statement Addressed to the Proprietors of Real Estate in the City and County of New York* (New York: J. Eastburn, 1818), 50.

47. Moore, 8.

48. On Moore's authorship of the poem, see Edwin G. Burrows and Mike Wallace, *Gotham: A History of New York City to 1898* (New York: Oxford University Press, 1999), 463.

49. Moore, *Plain Statement*, 49–50.

50. Moore, 49.

51. Moore, 54.

52. See Koeppel, *City on a Grid*, 137.

53. *Athenaeum*, June 16, 1849, 629.

54. Quoted in Koeppel, *City on a Grid*, 171.

55. Walt Whitman, "Letters from a Traveling Bachelor," *New York Sunday Dispatch*, November 25, 1849, quoted in Koeppel, *City on a Grid*, 172.

56. William Alexander Duer, *New York as It Was, during the Latter Part of the Last Century* (New York: Stanford and Swords, 1849), 18, quoted in Koeppel, *City on a Grid*, 172.

57. Henry James, "A Return to New York," *Harper's Monthly Magazine* 112 (1906): 900.

58. Lewis Mumford, *Sticks and Stones: A Study of American Architecture and Civilization* (New York: Boni and Liveright, 1924), 68; Edith Wharton, *A Backward Glance* (New York: D. Appleton-Century, 1934), 50.

59. John W. Reps, *The Making of Urban America* (Princeton, NJ: Princeton University Press, 1965), 99. For others who argue that real estate speculators were behind the adoption of the grid, see Edward K. Spann, "The Greatest Grid: The New York Plan of 1811," in *Two Centuries of American Planning*, ed. Daniel Schaffer (Baltimore: Johns Hopkins University Press, 1988), 11–40; Peter Marcuse, "The Grid as City Plan: New York City and Laissez-Faire Planning in the Nineteenth Century," *Planning Perspectives* 2 (1987): 287–310.

60. Koeppel, *City on a Grid*, 152–53.

61. See, for example, Elizabeth Blackmar, *Manhattan for Rent* (Ithaca, NY: Cornell University Press, 1989), 95–98.

62. Gouverneur Morris, *Notes on the United States of America* (Philadelphia: Office of the US Gazette, 1806), 11, 44.

63. Koeppel, *City on a Grid*, 115.

64. See Holloway, *Measure of Manhattan*, 55.

65. Randel's map of the Albany Post Road from Albany to Manhattan is housed today in the Albany County Hall of Records in Albany, New York.

66. Lewis Mumford, *The City in History: Its Origins, Its Transformations, and Its Prospects* (New York: Harcourt, Brace, and World, 1961), 421, quoted in Reuben S. Rose-Redwood, "Mythologies of the Grid in the Empire City, 1811–2011," *Geographical Review* 101, no. 3 (July 2011): 396–413, p. 398. For a discussion of the long-standing tradition that views the Commissioners' Plan as the expression of crude commercial interests, see Rose-Redwood, "Mythologies of the Grid," 398–401.

67. The text of the commissioners' *Remarks* can be found in Ballon, *Greatest Grid*, 40–41. The quotes are from p. 40.

68. Morris, *Notes*.

69. Morris, 44.

70. Morris, 12, 21.

71. "The Commissioners' Remarks," in Ballon, *Greatest Grid*, 40–41, p. 40.

72. Simeon De Witt, *The Elements of Perspective* (Albany, NY: H. C. Southwick, 1813), ix. For a discussion of De Witt's treatise on perspective and its significance for understanding the New York grid, see Rose-Redwood, "Mythologies of the Grid," 405–9.

73. De Witt, *Elements of Perspective*, ix.

74. De Witt, vii, xix, xvii.

75. Reuben Rose-Redwood suggests that, suspicious as they were of mass democracy, the commissioners viewed the straight streets and right angles of the grid as a way to discipline the unruly lower classes. See Reuben Skye Rose-Redwood, "Rationalizing the Landscape: Superimposing the Grid upon the Island of Manhattan" (master's thesis, Pennsylvania State University Department of Geography, 2002).

76. On Nancy Randolph, her marriage to Gouverneur Morris, and his relatives' chagrin, see Brookhiser, *Gentleman Revolutionary*, 180–85.

77. "The idea of a French republick was no doubt ridiculous, and the attempts fruitful in abominations," he wrote in 1806. See Morris, *Notes*, 46.

78. Quoted in Spiro Kostoff, *The City Shaped: Urban Patterns and Meanings through History* (Boston: Bulfinch, 1991), 102.

Chapter 6

1. Gerrard Koeppel, *City on a Grid: How New York Became New York* (New York: da Capo, 2015), 149.

2. Koeppel, 145–46.

3. Koeppel, 184.

4. On the Manhattan system of opening streets, assessing property, leveling and grading the landscape, and moving houses, see Hilary Ballon, ed., *The Greatest Grid: The Master Plan of Manhattan, 1811–2011* (New York: Museum of the City of New

York and Columbia University Press, 2012), 73–84. The estimate that 40 percent of the buildings in the city stood in the way of new streets and would therefore have to be demolished or moved is from Reuben Skye Rose-Redwood, "Rationalizing the Landscape: Superimposing the Grid upon the Island of Manhattan" (master's thesis, Pennsylvania State University Department of Geography, 2002), 108.

5. George Templeton Strong, diary entry for October 22, 1867, quoted in Ballon, *Greatest Grid*, 73.

6. See the Dripps Map (1850) and the Galt-Hoy Map (1879) in Robert T. Augustyn and Paul E. Cohen, eds., *Manhattan in Maps, 1527–1024* (New York: Dover, 2014), 110, 129.

7. Frederick Law Olmsted to Board of Commissioners of the Central Park, May 31, 1858, quoted in Koeppel, *City on a Grid*, 175.

8. William Irving, James, Kirk Paulding, and Washington Irving, *Salmagundi or the Whim-Whams and Opinions of Launcelot Langstaff, Esq., and Others*, rev. ed. (New York, 1860), 204–5. The original was published in the *Salmagundi* satirical magazine on May 16, 1807.

9. On Cooper, see quotes in Wade Graham, *American Eden* (New York: Harper Perennial, 2013), 60; Henry Hope Reed, *Central Park: A History and a Guide* (New York: Clarkson N. Potter, 1967), 3.

10. Walt Whitman, *New York Sunday Dispatch*, November 25, 1849, quoted in Koeppel, *Greatest Grid*, 172. See chap. 5 of this book.

11. William Cullen Bryant, "A New Public Park," *New York Evening Post*, July 3, 1844, quoted in Graham, *American Eden*, 87.

12. Andrew Jackson Downing, "The New York Park," *Horticulturalist and Journal of Rural Art and Rural Taste*, August 1, 1851, quoted in Graham, *American Eden*, 90.

13. On the attitudes of Washington Irving, James Fenimore Cooper, William Cullen Bryant, and Thomas Cole toward the American wilderness and the ways it shaped their work, see Roderick Frazier Nash, *Wilderness and the American Mind* (New Haven, CT: Yale University Press, 2014), first published 1967, pp. 72–83.

14. Downing, "New York Park."

15. On the consternation caused in the United States by the failure of the revolutions of 1848–49, the shock of mass immigration, and the impact they had on American intellectuals and national attitudes, see John Higham, *From Boundlessness to Consolidation* (Ann Arbor, MI: Clements Library, 1969).

16. Abraham Lincoln, Gettysburg Address.

17. Downing, "New York Park."

18. The roots of American fascination with wild nature, and hence with naturalistic landscapes both within and without cities, has been the topic of some debate. Roderick Nash has argued that in the absence of a long shared history early Americans looked to the vast wilderness of their country as a source of national identity and pride. Nevertheless, it was only with the rise of transcendentalism in the mid-nineteenth century that Americans came to value wilderness in its own right, as a fount of morality and wisdom. More recently, Richard Judd has argued

that the roots of Americans' environmental awareness go back further, to the work of eighteenth-century naturalists. But despite extending the timeline of American proto-environmentalism, Judd does not contest the mainstream view that it was the intervention of George Perkins Marsh, Ralph Waldo Emerson, Henry David Thoreau, Frederick Law Olmsted, and their associates in the middle of the nineteenth century that made wilderness conservation a major concern for American elites. See Nash, *Wilderness and the American Mind*; Nash, "The Cultural Significance of the American Wilderness," in *Wilderness and the Quality of Life*, Maxine E. McCloskey and James P. Gilligan (San Francisco: Sierra Club, 1969), 66–73; Richard W. Judd, *The Untilled Garden: Natural History and the Spirit of Conservation in America, 1740–1840* (New York: Cambridge University Press, 2009).

19. On the rise of English public parks, see Graham, *American Eden*, 86–87.

20. On Le Nôtre's designs for Greenwich Park, see F. Hamilton Hazlehurst, *Gardens of Illusion: The Genius of André le Nostre* (Nashville: Vanderbilt University Press, 1980), 3, 377–79.

21. On the spread of the French formal style in England in the late seventeenth and early eighteenth centuries see, Ralph Dutton, *The English Garden*, 2nd ed. (London: B. T. Batsford, 1947), 56–59.

22. On the origins of formal geometrical gardens and their use by French kings to present their rule as an outgrowth of the deep order of the universe, see Amir Alexander, *Proof! How the World Became Geometrical* (New York: Farrar, Straus, and Giroux, 2019), 69–183. On the adoption of French formal gardens by European royals as the emblem of monarchy, see Alexander, 187–209.

23. John Milton, *Paradise Lost* (Durham, NC: Duke Classics, 2012) book 4, 86, 89.

24. Joseph Addison, *Spectator*, June 25, 1712.

25. Alexander Pope, *Guardian*, September 29, 1713.

26. See *Spectator*, June 25, 1712.

27. On Stowe before Capability Brown's intervention, see Dutton, *English Garden*, 78–79. On Brown's redesign of Stowe, see Dutton, 88–89.

28. On Capability Brown and his many commissions throughout England, see Dutton, 88–91; Bill Bryson, *At Home: A Short History of Private Life* (New York: Anchor Books, 2010), 309–12.

29. Jean-Jacques Rousseau, *Julie, or the New Heloise*, in P. Stewart & J. Vaché, eds., *The Collected Writings of Rousseau* (Hanover: University Press of New England, 1997), 394.

30. Thomas Jefferson to John Page, May 4, 1786. On Thomas Jefferson's and John Adams's tour of English gardens, see Andrea Wulf, *Founding Gardeners: The Revolutionary Generation, Nature, and the Shaping of the American Nation* (New York: Vintage Books, 2012), 35–57.

31. The American founders, including Washington, Jefferson, and Madison, were all enthusiastic advocates of English-style gardening. Though no friends of England, they viewed the style much as its originators did, as a proper counterweight to tyranny—whether that of Louis XIV for the English or George III for the Americans. Over the following decades, they recast their own estates of Mount

Vernon, Monticello, and Montpellier in the rustic naturalistic style pioneered by Capability Brown. See Wulf, *Founding Gardeners*.

32. On the American reformers' hopes and intentions in creating the English garden that is Central Park, see Geoffrey Blodgett, "Landscape Design as Conservative Reform," in *Art of the Olmsted Landscape*, ed. Bruce Kelly, Gail Travis Guillet, and Mary Ellen W. Herne (New York: New York City Landmarks Preservation Commission and Arts Publisher, 1981), 111–23.

33. Andrew Jackson Downing, "The New York Park," *Horticulturalist and Journal of Rural Art and Rural Taste*, August 1, 1851.

34. On the fight for "the New York Park," see Reed, *Central Park*, 14–17; Ian R. Stewart, "The Fight for Central Park," in Kelly, Guillet, and Hern, *Art of the Olmsted Landscape*, 87–97.

35. Frederick Law Olmsted, *Walks and Talks of an American Farmer in England* (New York: George P. Putnam, 1852), 79.

36. Olmsted, *Walks and Talks*, 81.

37. On Frederick Law Olmsted's complicated path to becoming a landscape designer, see Graham, *American Eden*, 91–93; Reed, *Central Park*, 21–22.

38. Frederick Law Olmsted, *Journeys and Explorations in the Cotton Kingdom* (London: Sampson Low, Son, 1861).

39. Quoted in Graham, *American Eden*, 93.

40. Egbert Viele, *First Annual Report on the Improvement of the Central Park* (New York: McGrath Publishing Co., 1857), 12.

41. Egbert Viele, "Topography of New York and Its Park System," in *The Memorial History of the City of New York*, ed. James Grant Wilson, 5 vols. (New York: New York History, 1893), 4:556–57, quoted in Roy Rosenzweig and Elizabeth Blackmar, *The Park and the People* (Ithaca, NY: Cornell University Press, 1992), 64.

42. On the population of the park-to-be and Seneca Village, see Rosenzweig and Blackmar, *Park and the People*, 63–73.

43. For the condition of the grounds of the future Central Park, as well as for the quotes, see Reed, *Central Park*, 19–20. On the Park Keepers, see Reed, 35.

44. On Vaux's partnership with Downing and his advocacy for an open design competition, see Graham, *American Eden*, 88–93; Melvin Kalfus, *Frederick Law Olmsted: The Passion of a Public Artist* (New York: New York University Press, 1990), 193.

45. On Olmsted and Viele, see Reed, *Central Park*, 21–22.

46. Olmsted, *Walks and Talks*, 79.

47. The quote is from Kalfus, *Frederick Law Olmsted*, 194.

48. Frederick Law Olmsted, "The Misfortunes of New York," in *Civilizing American Cities: A Selection of Frederick Law Olmsted's Writings on City Landscapes*, ed. S. B. Sutton (Cambridge, MA: MIT Press, 1979), 43–51, p. 45, excerpted from City of New York, Document no. 72 of the Board of the Department of Public Parks, 1877.

49. Frederick Law Olmsted and Calvert Vaux, *Description of a Plan for the Improvement of the Central Park: "Greensward"* (1858; repr., New York: Aldine, 1868), 5.

50. Frederick Law Olmsted, "The Justifying Value of a Public Park," *Journal of Social Science, Containing the Transactions of the American Association* 12 (December 1880): 147–64, pp. 163–64.

51. The quote is from a report by Olmsted published in the *Second Annual Report of the Commissioners of the Central Park*, January 1, 1858, quoted in Stephen Rettig, "Influences across the Water: Olmsted and England," in Kelly, Guillet, and Hern, *Art of the Olmsted Landscape*, 79–85, p. 79.

52. Olmsted, Vaux, and Co., "Report of the Landscape Architects," in *Sixth Annual Report of the Commissioners of Prospect Park*, 1866, in *Annual Reports of the Brooklyn Park Commissioners, 1861–1873* (Brooklyn: Board of Commissioners of Prospect Park, 1873), 91–117, pp. 93, 95.

53. On the scale of the work and the workforce, see Henry Hope Reed, *Central Park: A History and a Guide* (New York: Clarkson N. Potter, 1967), 30–31; Anna Maria Gillis, "Public Parking," *Humanities: The Magazine of the National Endowment for the Humanities* 33, no. 1 (January/February 2012): 28–33, p. 31.

54. Frederick Law Olmsted, "Public Parks and the Enlargement of Towns," *Journal of Social Science, Containing the Transactions of the American Association*, no. 3 (1871): 1–36, pp. 12, 20–22. Olmsted gave his address at the Lowell Institute in Boston on February 25, 1870.

55. Frederick Law Olmsted, "Public Parks and the Enlargement of Towns," 34.

56. On colonial gridded towns and cities in the United States, see John W. Reps, *The Making of Urban America: A History of City Planning in the United States* (Princeton, NJ: Princeton University Press, 1965), 128–29, 157–82.

57. On Woodward's design for Detroit and its aftermath, as well as on Indianapolis and Madison, see Reps, *Urban America*, 264–93.

58. On Olmsted's role in the design of Stanford University, see Albert Fein, "The Olmsted Renaissance: A Search for National Purpose," in Kelly, Guillet, and Hern, *Art of the Olmsted Landscape*, 99–109, pp. 103–4.

59. On Olmsted's design for Riverside, Illinois, see Reps, *Urban America*, 348.

60. The commenter was responding to Olmsted's 1883 proposal for the town of Tacoma, Washington. The plan, in this case, was rejected. See Reps, 410.

61. Quoted in Reps, 344.

62. A partial list of Olmsted-designed suburbs can be found in Wade Graham, *American Eden* (New York: Harper Perennial, 2011), 102.

63. Quoted in Graham, 102.

64. Quoted in Reed, *Central Park*, 35.

65. Graham, *American Eden*, 95.

66. On Olmsted's involvement in the Yosemite land grant of 1864, see Nash, *Wilderness and the American Mind*, 106–7; Graham, *American Eden*, 95–96; Melvin Kalfus, *Frederick Law Olmsted: The Passion of a Public Artist* (New York: New York University Press, 1990), 247–48. In an 1865 report to the California legislature, Olmsted supported the preservation of Yosemite Valley using the same arguments that had secured the creation of Central Park in New York: that access to natural beauty is es-

sential for men and women's mental and moral health. See Olmsted, *Yosemite and the Mariposa Grove: A Preliminary Report*, Yosemite Online, https://www.yosemite.ca.us/library/olmsted/report.html, reprinted with forewords by Dayton Duncan and Ken Burns (San Francisco: Yosemite Conservancy, 1995).

67. For a summary of the emergence of the National Park System, see White, Richard White, *It's Your Misfortune and None of My Own: A History of the American West* (Norman: University of Oklahoma Press, 1991), 409–415.

68. For more on Olmsted's role in launching the national parks movement, see Dennis Drabelle, *The Power of Scenery: Frederick Law Olmsted and the Origin of National Parks* (Lincoln, NE: Bison Books, 2021).

69. On the different varieties of the American grid, including different styles of street naming and numbering, see Reuben Rose-Redwood and Lisa Kadonaga, "The Corner of Avenue A and Twenty-Third Street: Geographies of Street Numbering in the United States," *Professional Geographer* 68, no. 1 (2016): 39–52.

Conclusion

1. Ashley Fantz, "Oregon Standoff: What the Armed Group Wants and Why," CNN, January 4, 2016, http://www.cnn.com/2016/01/04/us/oregon-wildlife-refuge-what-bundy-wants/.

2. Corky Siemaszko, "Meet Ammon and Ryan Bundy, the Activists Leading the Oregon Standoff," NBC News, January 4, 2016, http://www.nbcnews.com/news/us-news/meet-ammon-ryan-bundy-activists-leading-oregon-standoff-n489766.

3. Siemaszko, "Meet Ammon and Ryan Bundy."

4. On the occupiers' religious affiliations and affinity to the "Sovereign Citizen" movement, see the Wikipedia article on "Citizens for Constitutional Freedom," accessed October 13, 2022, https://en.wikipedia.org/wiki/Citizens_for_Constitutional_Freedom#Known_members.

5. For a list of the group's known members and their residences, see the Wikipedia article on "Citizens for Constitutional Freedom," accessed August 24, 2022, https://en.wikipedia.org/wiki/Citizens_for_Constitutional_Freedom#Known_members.

6. For a beautiful and poetic account of a similar conflict taking place on ancestral Apache lands in southeast Arizona, see Lauren Redniss, *Oak Flat: A Fight for Sacred Land in the American West* (New York: Random House, 2020).

7. In recent decades, historians have turned their critical attention to the troubled history of the West, amply demonstrating that many roads led to the current confrontations. Environmentalist historians such as William Cronon, Donald Worster, and Ted Steinberg have emphasized the power of capitalism to disrupt and reshape western lands. Worster and Donald Pisani have also paid particular attention to the ways in which water scarcity and need for hydrological controls have led to the disproportionate concentration of power in government hands. Western historians such as Patricia Nelson Limerick and Benjamin Madley have pointed to

the long history of warfare, dispossession and even genocide carried out against Native Americans, and Greg Grandin has argued that that the West was and remains a fount of racism and violence. See William Cronon, *Nature's Metropolis: Chicago and the Great West* (New York: W.W. Norton, 1991); Donald Worster, *Under Western Skies: Nature and History in the American West* (Oxford: Oxford University Press, 1992); Ted Steinberg, *Down to Earth: Nature's Role in American History* (Oxford: Oxford University Press, 2002); Donald J. Pisani, *Water, Land, and Law in the West* (Lawrence: University Press of Kansas, 1996); Patricia Nelson Limerick, *The Legacy of Conquest : The Unbroken Past of the American West* (New York: W.W. Norton, 1987); Benjamin Madley, *An American Genocide: The United States and the California Indian Catastrophe, 1846–1873* (New Haven, CT: Yale University Press, 2016); and Greg Grandin, *The End of the Myth: From the Frontier to the Border Wall in the Mind of America* (New York: Metropolitan Books, 2019). For an overview, see Richard White, *It's Your Misfortune and None of My Own: A New History of the American West* (Norman: University of Oklahoma Press, 1991).

8. For a full discussion of the political uses of geometry to buttress royal authority in Europe and imperial authority beyond, see Amir Alexander, *Proof! How the World Became Geometrical* (New York: Farrar, Straus, and Giroux, 2019).

Index